U0201343

第2版

舌尖上的英语

创想外语研发团队　编著

中国纺织出版社

图书在版编目（CIP）数据

舌尖上的英语 / 创想外语研发团队编著. —2版
. —北京：中国纺织出版社，2018.10（2021.8 重印）
ISBN 978-7-5180-4621-8

Ⅰ.①舌… Ⅱ.①创… Ⅲ.①饮食-文化-英语
Ⅳ.①TS971.2

中国版本图书馆CIP数据核字（2018）第014063号

责任编辑：汤　浩　　　　责任印制：储志伟

中国纺织出版社出版发行
地　　址：北京市朝阳区百子湾东里A407号楼　邮政编码：100124
销售电话：010—67004422　传真：010—87155801
http: //www. c–textilep. com
E–mail: faxing@c–textilep. com
中国纺织出版社天猫旗舰店
官方微博 http: //weibo. com/2119887771
佳兴达印刷（天津）有限公司印刷　各地新华书店经销
2015年2月第1版　2018年10月第2版　2021 年 8 月第 3 次印刷
开　　本：880×1230　1 / 32　印张：10
字　　数：300千字　定价：45.00元

前言

“民以食为天”，自古以来中国人就喜欢品尝美食，谈论美食。在国内，美食早已成为舌尖热门话题，同样，在国外，美食与天气一样是人们沟通的常用话题。对每一位英语学习者来说，或许谈论天气倒是轻而易举，信手拈来，但是与外国人聊起美食，好多人却苦于词汇匮乏而有口难开。国外菜式风味各不相同，还有不同的美食文化、餐桌礼仪等，令人眼花缭乱，学习也无从下手。

此时，你是否正在寻觅一本全面的舌尖美食英语书？资深专业外教团队给英语学习者们奉献了一份原汁原味的英语美食大餐。读者拿到这本书，学习英语就像烹饪美食一样，一步一步学，轻轻松松学，最后品尝最美的英语口语！

本书特色

1. 语言精美，英语更地道！

本书由我们的资深专业外教团队，秉承只做经典英语口语理念，倾力打造纯正、精美、有味道的美食英语口语。每一个对话片段，都是真实的美食英语场景，每一句话，都是经典口语句。

2. 专业时尚，学习更有效！

本书涵盖专业的餐饮词汇和语句，考究时尚的美食英语表达，让英语学习者交流起来更自信，更有效！

3. 美食文化大餐，学习更有品位！

本书介绍美、英、意、法等各大欧美及韩日菜式文化，从开胃品，到主菜、配菜，再到甜品，应有尽有，品味时尚美食文化，做有品位的英语潮人！

4. 内容编排有序，学习有章可循！

本书按照饮食习惯，从食材及厨具购买，到欧美菜式、韩日菜式，再到用餐习惯、用餐礼仪等顺序编排，让读者学习有章可循，学习有步骤！

本书内容

本书基本涵盖美食烹饪的各个环节，包括欧美各大菜式、韩日菜式等各大国外美食英语用语。本书分为9部分，共47个话题，具体包括饮食习惯、购买食材及厨具、美国美食、英伦风味、意式美味、法国情调、日本料理、韩式料理、开始用餐。

本书每个话题，具体内容编排如下：

1. 舌尖美食词汇

该部分是常用的、鲜活的词汇表达。为读者积累想说就说的美食词汇。

2. 舌尖美食句

该部分是精心选取的经典美食英语语句，简约而精炼，需要你大胆说出口。

3. 舌尖聊美食

该部分每个场景都由两个对话构成，每个场景对话都是美食烹饪的真实场景再现，以教授地道纯正的口语，品味各国美食文化。“开胃词组”和“鲜味单词”分别对场景对话中的重点短语和词汇做了解析，让你在学习地道口语的同时，夯实基础知识，提升英语口语从细节开始！

4. 舌尖美食文化

该部分涉及欧美、韩日美食时尚文化，不但可以让读者了解东西方美食文化的差异，避免跨文化交流中的障碍，而且可以让读者提升自己的文化涵养。

编者

2018年3月

Part 1 Eating Habits 饮食习惯

Part 2 Buying Ingredients and Kitchenware 购买食材及厨具 033

Part 3 American Cuisine 美国美食 073

Part 6 French Cuisine 法国情调 177

读书笔记

Part 1

Eating Habits
饮食习惯

UNIT 01

Vegetarians 素食主义

舌尖美食词汇

veggie 素食者	**vegetarian meal** 素食餐
vegetarian diet 素食	**veggie burger** 素食汉堡
demi-veg 半素食者	**vegetarian restaurant** 素食餐厅
vitamin 维生素	**energy** 能量
veggie pasta 蔬菜意大利面	**pescetarian** 只吃鱼和素食的人
lacto-ovovegetarians 乳蛋素食者	**vegan** 严格素食主义者
nutrition 营养	**protein** 蛋白质
egg 鸡蛋	**milk** 牛奶

舌尖美食句

1. I'm a veggie. 我是素食者。

2. I need vegetarian meals. 我吃素餐。

3. I didn't know you are a vegetarian! 我不知道你是素食主义者！

4. She just told me she's a vegetarian! 她刚告诉我她是一位素食主义者！

5. Eating a vegetarian diet is one way to keep you healthy.	吃素是一种保持身体健康的饮食方式。
6. I've been vegan for a long time.	我坚持严格吃素已经很久了。
7. Could we go to a vegetarian restaurant for dinner?	我们去一家素食餐厅吃晚饭，怎样？
8. I am allergic to meat.	我对肉过敏。
9. Do you know any famous vegetarians?	你知道哪些著名的素食主义者吗？
10. He is a demi-veg who eats fish but not meat.	他是一个吃鱼但不吃肉的半素食者。
11. Could you try vegetarianism for a month?	你尝试过做一个月的素食者吗？
12. Have you heard of veggie burger?	你听说过素食汉堡吗？
13. Many restaurants offer some veggie options.	许多饭馆提供素菜。

舌尖聊美食

Conversation 1

I'm starving! How about going out to grab something to eat?	我饿死了！要不要出去吃点东西？
Sure! What would you like to eat?	好啊！你想吃什么？
I really feel like having a big juicy steak!	我好想来一块美味多汁的牛排啊！

Well, I don't eat meat, but that's fine, I am sure wherever we are going they will have other options, right?

我是不吃肉的。不过没关系。我们去的地方肯定有很多种选择。

I didn't know you are a vegetarian!

你是素食主义者呀？我还不知道呢。

I'm not, I am a vegan.

不，我是一个严格素食主义者。

A what?

那是什么意思？

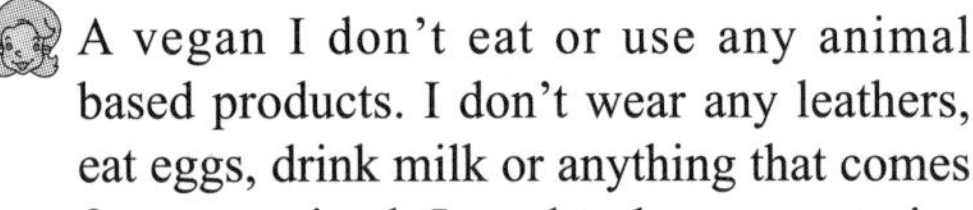
A vegan I don't eat or use any animal based products. I don't wear any leathers, eat eggs, drink milk or anything that comes from an animal. I used to be a pescetarian before, which basically means you don't eat meat, but still have fish and seafood.

意思就是我不食用任何来自动物身上的东西。不穿皮草，不食用蛋类、奶制品或其他任何来自动物的食品。我以前是鱼素者，就是不吃肉但是可以吃鱼类和海产。

Wow! How interesting! It must be tough!

哇！真有意思！肯定挺难的吧。

It's a bit difficult to find vegetarian friendly restaurants sometimes, but since more and more people are vegetarians or vegans nowadays, it's getting better.

有时候确实很难找到一家支持素食主义的餐馆。不过现在越来越多的人成为素食主义者或是严格素食主义者，情况也就好多了。

开胃词组

feel like 想要……

come from 来自……

used to 过去常常

it's difficult to 很难（做）……

pescetarian 鱼素者（吃鱼不吃肉的素食者）

鲜美单词

starving adj. 饥饿的，挨饿的

grab v. 取，抓

juicy adj. 多汁的，有趣的

option n. 选择

vegetarian adj. & n. 素食的，素食主义的；素食者

restaurant n. 饭店，餐馆

vegan n. 严格的素食主义者

Conversation 2

I'm hungry, let's go eat.	我饿了，咱们去吃东西吧。
Okay, some place with good vegetarian things, though.	好吧。不过，得去一个提供美味素食的地方。
I got it, you're a vegan. But what is the reason?	没错，你是素食主义者。你为什么吃素？
I don't like cruelty to animals.	我不喜欢对待动物的残酷行为。
Well, I don't like that either, but I still like meat.	嗯，我也不喜欢那样，但我仍然喜欢吃肉。
Also, I just don't really like the taste of meat.	另外，我只是不喜欢肉的味道。
Really? Even no chicken or beef or pork?	真的吗？不喜欢鸡肉、牛肉还是猪肉？
Nope.	都不喜欢。
What about bacon?	培根呢？

Absolutely not.	完全不喜欢。
What? Everyone likes bacon.	什么？大家都喜欢培根。
Not me. I can't stand it.	不包括我。我受不了它的味儿。
Wow. I don't think I could live without bacon.	哇。我觉得没有培根我都活不下去。
It's not that hard! Just stop eating meat. Come on, tonight we both eat vegetarian food.	没有那么难吧！那就别吃肉了。走吧，今晚我们一起吃素。
Alright, I'll try it. Just for you, buddy.	好吧，我试试吧。可都是为了你，伙计。

开胃词组

vegetarian thing 素食的东西

the taste of ……的味道

cruelty to animals 虐待动物，残害动物

stop doing sth. 停止做……

鲜美单词

reason n. 原因，理由

cruelty n. 残暴，残忍

taste n. 味道，口味

beef n. 牛肉

pork n. 猪肉

bacon n. 培根

素食者（vegetarian），指不食用肉类，鱼类，禽类及其副产品的人，素食者容易出现营养不良症状，需要注意补充营养。素食者根据避免动物制品的程度，可分为以下几种：

严格素食者（strict vegetarian/vegan），是指不食用动物的肉，包括肉类、禽类、鱼类，也不食用来自动物身体的物品，比如蛋类、奶类。

半素食主义（semi-vegetarianism），是指可能基于健康、道德或信仰而不食用牛、羊、猪等哺乳动物红肉的素食主义者。这些人会食用部分禽类和海鲜。半素食主义包括弹性素食主义（flexitarianism）。

斋食（buddhist vegetarianism），是指避免食用所有由动物制成的食品，也不食用青葱、大蒜、洋葱、韭菜等在内的葱属植物。

奶素主义（lacto vegetarianism），是指食用奶类相关产品的素食主义者，如奶酪、奶油或酸奶，而不食用蛋及蛋制品。

奶蛋素食主义（lacto-ovo vegetarianism），是指只食用奶蛋类食品来取得身体所需之蛋白质，如蛋和奶类食品，不吃肉类或含肉类制品的素食主义者。

蛋素主义（ovo vegetarianism），这类素食主义者食用蛋类产品，而不吃奶及奶制品。

生素食主义（raw foodism），是指将所有食物保持在天然状态，即使加热也不超过47℃的素食主义者。

鱼素主义（pescetarianism），是指以素食为主，偶尔会食用鱼和其他水产品的素食主义者。

UNIT 02

On Diet
节食减肥

舌尖美食词汇

overweight 超重	**on diet** 节食
shed a few pounds 甩掉几磅	**eating disorder** 饮食混乱
take more exercise 多做运动	**out of shape** 身材变形
skinny enough 够瘦的	**lose weight** 减肥
fat 脂肪	**weight loss plan** 减肥计划
physical fitness 体能健康	

舌尖美食句

1. I'm on a diet.	我正在节食减肥。
2. She's overweight.	她超重了。
3. I think I'm out of shape.	我觉得我身材走样了。
4. I'm just trying to shed a few pounds in the natural way.	我只是想以自然的方式甩掉几磅。
5. What's the secret diet?	你节食的秘方是什么？

6. You need to reduce the amount of fat in your diet.	你要减少饮食当中的脂肪摄取量。
7. Are you going on a diet?	你在节食吗？
8. I hope you can get slimmer in a healthy way.	我希望你以健康的方式来减肥。
9. He is on a diet to reduce some weight.	他正在节食以减轻体重。
10. You're skinny enough without going on a diet!	你不必节食就已经够瘦的了！
11. I am a little overweight. I should go on a diet.	我有点超重了，得节食才行。
12. I suggest you take more exercise.	我建议你多做运动。

舌尖聊美食

Conversation 1

Did you have breakfast this morning?	今天你吃早饭了吗？
No. I'm on a diet. I would like to shed a few pounds of weights.	没有，我在节食。我想甩掉几磅肉。
I don't think skipping meals is the best way to go if you want to lose weight.	我不觉得节食是减肥的好方法。
How can I lose weight if I don't eat less?	我不吃少点，那还怎么减肥呢？

Well, without normal meals, your body will try to store more fat and calories than usual because skipping meal lowers your metabolism. You will eat a lot more for lunch if you don't have enough to eat in the morning.

嗯，不吃正餐，你的新陈代谢就会变慢，你的身体就会比平常囤积更多的脂肪和热量。如果不吃早餐，你午餐就会吃得更多。

I'm always wondering how you can stay in such a good shape. You don't even go to the gym, and your diet is normal. What's your secret?

我想知道你是如何保持这么好的身材的。你不去健身房，而且想吃就吃。你有什么秘方吗？

Well, a balanced diet is the key. I try to make sure I always have enough protein and vegetables and less fat and sugar for each meal.

嗯，平衡饮食是关键。我总是确保摄入足够的蛋白质和蔬菜，每餐少吃含脂肪和糖的食物。

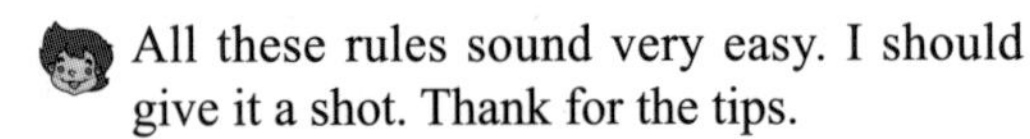

All these rules sound very easy. I should give it a shot. Thank for the tips.

这些原则听起来很简单。我应该试试。谢谢你的建议。

开胃词组

have breakfast 吃早餐

be on a diet 节食

shed a few pounds 甩掉几磅肉

lose weight 减肥

give it a shot 试一试

鲜美单词

breakfast *n.* 早餐，早饭

shed *n.* 摆脱，去除

pound *n.* 磅

weight *n.* 重量，体重

skip *v.* 跳跃，跳过

calorie *n.* 卡，卡路里
metabolism *n.* 新陈代谢
secret *n.* 秘密，机密
protein *n.* 蛋白，蛋白质
shot *n.* 尝试，努力

Conversation 2

Would you like to have some ice cream? I've got a variety of flavors for you to choose from. I've got strawberry, peach, chocolate, chocolate chip, chocolate brownie, coffee, vanilla, rocky road, butter pecan and praline.	你想来点冰淇淋吗？我有许多种口味可供选择。有草莓味的、鲜桃味的、巧克力味的、巧克力片的、核仁巧克力的、咖啡的、香草味的、石板街坚果巧克力的、奶油山核桃味的、果仁糖味的。
Wow! So many choices you have! I wish I could, but I can't. I'm on a diet to lose my weight.	哇!那么多！我想吃啊，但是我在节食减肥。
Come on, it's just a bite.It doesn't really matter if you just have a bite.	尝一口没什么大碍的。
Piss off! Don't lure me into it.Please!	走开。请不要诱惑我。
Gee! You are really stubborn.	哎呀！你真固执。
You're right. I'm not so easily coaxed into doing something that I think it is wrong.	没错，错误的事情，我不会轻易被引诱的。
Well, I'd better give up teasing you. Otherwise, you will be addict to ice cream again.	嗯，我最好不要再取笑你了。否则，你会再次迷上冰淇淋的。

开胃词组

a variety of 各种各样

have a bite 咬一口，尝一口

piss off 滚开，走开

lure sb. into 引诱某人做某事

coax into 劝诱，哄

give up 放弃

ask for 要求

鲜美单词

flavor *n.* 味，味道

strawberry *n.* 草莓

brownie *n.* 核仁巧克力

vanilla *adj.* 香草的，香草味的

pecan *n.* 山核桃

lure *v.* 引诱，诱惑

stubborn *adj.* 顽固的，固执的

coax *v.* 哄，劝诱

otherwise *conj.* 否则，不然

addict *n.* 有瘾的人，入迷的人

舌尖美食文化

Being On a Diet

节食

我们想拥有完美身材，使尽浑身解数减肥，有的采取节食法，通过减少进食来达到减肥目的。可你听说过5比2节食法吗？

The 5:2 diet, also written as 5/2 diet, is a fad diet which involves severe calorie restriction for two non-consecutive days a week and normal eating the other five days, which originated and became popular in the UK, and spread in Europe and to the US. It is a form of intermittent fasting. The diet is claimed to promote weight loss and to have several beneficial effects on health. According to the UK National Health Service there is limited evidence on the safety and effectiveness of the diet; they advise people considering it to consult their doctor.

5比2节食法是一种时兴的节食方法，指一周7天，2天严格控制摄入卡路里，其余5天正常进食，控制热量的2天不是连续的。这种节食方法源自英国并在那里广受欢迎，之后传到了欧洲和美国。这是间断性节食的一种方式，据说能够帮助减肥，还对健康很有益处。不过，英国国民健康服务机构表示，这种节食方法在安全和有效性方面的证据很有限，建议人们尝试之前先咨询医生。

Men may eat 600 calories on fasting days, and women 500. A typical fasting day may include a breakfast of 300 calories, such as two scrambled eggs with ham, water, green tea, or black coffee, and a lunch or dinner of grilled fish or meat with vegetables, amounting to 300 calories. The daily 500 or 600 calorie limit requires small portions.

在节食日，男性可以摄入600卡的热量，女性可摄入500卡。节食日一天的餐食大概包括：早餐300卡，比如2个炒鸡蛋加火腿、水、绿茶或者黑咖啡；午餐或晚餐300卡，可以吃烤鱼或烤肉，再加点蔬菜。一天500或600卡的热量意味着每一餐分量都很小。

UNIT 03 Organic Food 有机食物

舌尖美食词汇

organic food 有机食物	**skimmed milk** 脱脂牛奶
organic carrot 有机胡萝卜	**organic apples** 有机苹果
without pesticides or fertilizers 无农药无化肥	**so tasty** 那么美味
shop for 购买	**bean sprout** 豆芽
onion 洋葱	**green food** 绿色食品
pesticide residues 农药残留	**nutrient** 营养物质

舌尖美食句

1. Do you like organic food?	你喜欢有机食物吗?
2. I usually eat organic food, skimmed milk, fresh fish and organic carrots.	我通常吃全麦面包，脱脂牛奶，鲜鱼和有机胡萝卜。
3. Do you have any organic apples?	你这里卖有机苹果吗?
4. They have a wide range of organic food.	他们有许多有机食品出售。
5. Is organic food safe?	有机食品安全吗?

6. Organic food is grown without using pesticides or fertilizers.	有机食品是不用杀虫剂和化肥来种植的。
7. Healthy organic food can be so tasty!	健康有机食品是美味的。
8. People begin to shop for organic food.	人们开始喜欢购买有机食品。
9. The demand for organic food is increasing.	有机食品的需求增多了。
10. What is the difference between organic food and other food?	有机食品与其他食品有何不同？
11. He works on an organic farm.	他在一个有机食品农场工作。

舌尖聊美食

Conversation 1

How is the soup?	这汤喝起来怎么样？
Well, it isn't as bad as I thought. Could you show me how to make it next time?	出乎意料，不错！你下次能教我做吗？
Sure. I'm glad you like it. I'll write down the recipe for you later on. Just make sure you get all the ingredients from the organic food store.	当然可以！很高兴你喜欢。我等会儿给你写菜谱。只不过你得先去菜市场买有机食物材料。
What is organic food anyway? I often hear that name but I just don't know the exact meaning.	有机食物是什么？我常常听说这个名，就是不知道是什么意思。
It refers to the food that is grown without using pesticides or fertilizers. It's chemical-free and safe to eat. I shop organic food only.	它通常指的是不用杀虫剂或化肥种植的食物，不含化学物质，可安全食用。我只购买有机食物。

Yeah, that's good for health! But where can I get these organic foods?	这对身体很好。但我去哪里能买到这些有机食物?
Um, some large supermarkets have organic food section, you can have a check.	嗯，一些大超市设有有机食物专区，你看看吧。
And it's quite expensive, isn't it?	有机食物很贵吧?
Yeah. Things which are healthy and rare usually cost you more.	对，健康和稀有的东西都很贵的。
I agree with you.	没错。

开胃词组

as bad as 与……一样糟糕

next time 下次

write down 记下，写下

later on 稍后，以后

refer to 涉及，指的是

agree with 同意，赞同

鲜美单词

recipe *n.* 食谱，菜谱

ingredient *n.* （烹饪的）原料

organic *adj.* 有机的，绿色的，无农害的

pesticide *n.* 杀虫剂，农药

fertilizer *n.* 肥料，化肥

section *n.* 部分

expensive *adj.* 昂贵的

Conversation 2

Hey, Jerry. I heard that you have a garden.	喂，杰里，我听说你有一个花园。
Yeah, I got a vegetable garden in my backyard where I grow organic food.	对，我在后院搭建了个菜园，种植有机作物。
Oh, nice! Can you tell me about your organic plants and how you eat?	真好！你能跟我说说你家种的有机作物吗，你是怎么吃的？
Sure! Well, in the summer, I grow a lot of salad vegetables: lettuce, green onions, tomatoes, and over the winter, I grow a lot of carrots or cabbages, broccoli.	当然可以！夏天的时候我种许多能用来做沙拉的蔬菜：包括莴苣、大葱、西红柿；而冬天则种大量的胡萝卜、卷心菜和花椰菜。
Wow, you grow vegetables all year round!	哇，你一整年都种蔬菜！
Yes, that's how I got enough organic food for the whole year! I have never bought vegetables from the markets.	是的，我一整年都能吃到有机食物！我从不在市场买菜。
And what kind of food do you like to make with these vegetable?	你喜欢用这些蔬菜做什么菜来吃？
I just love the raw vegetables, so salads are great, and I eat them with Tofu or…yeah, vegetables are good. I like to make soup with fresh veggies as well.	我喜欢吃生蔬菜，做成沙拉会很棒，我会搭配豆腐来吃，或者……总之，蔬菜非常棒。我喜欢用新鲜的蔬菜做汤。
Oh, what kind of soup?	哪种汤？
Well, I usually don't follow the written recipes. I do it in my way. I throw any veggies into the pot. Of course, tomato soup and corn soup are amazing.	实际上我并不按照食谱做，用自己的方法来烹饪。我把所有的蔬菜都放进锅里。当然啦，西红柿汤和玉米汤非常棒。

Sounds good. How do you make the soups? I mean what is the procedure?

听起来不错。你怎么做汤？我是说步骤是什么？

OK, well, basically I choose whatever vegetables I have, chop them up and cook them, I like crunchy things so I boil them lightly. And then just add a bit of some salt, then a soup is done, a veggie soup.

好，基本上来说我选用我能找到的蔬菜做汤，把蔬菜切碎然后煮。我喜欢吃起来脆脆的蔬菜，所以稍微煮一下。再放点盐就行了，这样就做成了汤，蔬菜汤。

That's must be a very special soup! But for me, I can't eat without meat.

那汤肯定非常特别！但是对于我，没有肉我可吃不下。

I'm a vegetarian, you remember?

我是一名素食者，还记得吧？

Oh, yes, you are.

噢，对了，没错。

开胃词组

all year round 全年，一年到头

raw vegetables 生蔬菜

as well 也，还有

a bit of 一点儿

鲜美单词

garden n. 花园，菜园

backyard n. 后院

lettuce n. 莴笋，生菜

onion n. 洋葱，葱头

broccoli n. 花椰菜

raw adj. 生的，未经加工的

fresh adj. 新鲜的

pot *n.* 锅

procedure *n.* 工序，步骤，过程

chop *v.* 砍，剁，切

crunchy *adj.* （食物）硬脆的，松脆的

有机食物越来越受欢迎

More land around the world is being dedicated to organic farming. The World Watch Institute says since 1999 there's been a more than three-fold increase to 37 million hectares.

据世界观察研究所调查，世界上越来越多土地被用来种植有机作物，自从1999年来，有机作物种植面积增长了3倍，达到3700万公顷。

"Organic farming is farming without chemical inputs, like pesticides and fertilizers. Instead of using those inputs it uses a variety of natural techniques, like rotating crops and applying compost to fields – and growing crops that will return nutrients to the soil naturally instead of via chemicals." said World Watch researcher Laura Reynolds, who co-authored a new report on *the growth of organic agriculture*.

"有机农场种植不使用化学物质，如杀虫剂和化肥，而通过各项天然技术，比如轮换种植作物和使用混合肥料，从而让营养物质回归土壤。"《有机农业的发展》最新报告的合著者、世界观察所研究员劳拉·雷诺兹说道。

It said that a range of public health and environmental benefits.

有机农业同样有利于公共卫生和环境。

"It delivers fewer pesticides and chemicals to what we eat and to the farmers growing the food. It also delivers a range of economic benefits to farmers growing organically because they found they can get a much higher price if their food is certified organic." she said.

劳拉说："有机农业减少了人们食物当中的药剂和化学物质含量，也有利于增加种植者的收入，因为有机作物的价格会高许多。"

读书笔记

I Like to Eat Snacks 爱吃零食

舌尖美食词汇

some snacks 一些零食	**potato chips** 薯片
biscuit 饼干	**chocolate** 巧克力
midnight snack 夜宵	**milk shake** 奶昔
new flavour 新口味	**lollipop** 棒棒糖
cracker 薄脆饼干	**peanut** 花生
muffin 松饼	**cookie** 曲奇饼干
doughnut 甜甜圈	

舌尖美食句

1. I prefer to snacks rather than a full meal.
相对于正餐来说，我更喜欢吃零食。

2. I have potato chips that are in three new flavours.
我有三种新口味的薯片。

3. What snacks do you like best?
你最喜欢吃什么零食?

4. I'm going to buy some snacks.
我去买点儿零食。

5. I think she is indulging on unhealthy foods with crisps, biscuits, cakes and chocolates.	我认为她沉溺于薯片、饼干、蛋糕和巧克力等不健康的食物中。
6. I'm skipping dinner, but for a midnight snack, I'm having a milk shake.	我不吃晚饭了，但是夜宵的话我想喝点奶昔。
7. He started to nibble his biscuit.	他开始啃饼干了。
8. I'm crazy about snacks.	我特喜欢吃零食。
9. I snack between meals.	我吃饭之前会吃零食。
10. Would you like to try these fresh doughnuts?	你想尝一尝这些新鲜的甜甜圈吗？
11. My brother loves to eat crackers for his snack.	我弟弟喜欢把薄脆饼当零食吃。

舌尖聊美食

Conversation 1

Hi, Peter. Where have you been?	嘿，皮特。你去哪了？
I went to the supermarket. I've run out of snacks.	我去超市了。我的零食吃没了。
But you just brought back a pile of snacks.	但是你刚买回来一大堆零食啊。
Well. You know I'm crazy about snacks. I can't live without them.	嗯，你也知道我太喜欢吃零食了。没有零食我可活不了。

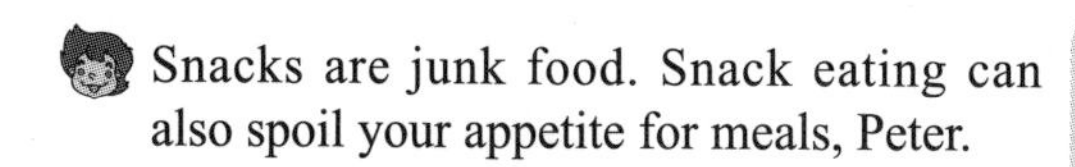

Snacks are junk food. Snack eating can also spoil your appetite for meals, Peter.

皮特，零食是垃圾食品。吃太多零食，你就会没有胃口吃正餐。

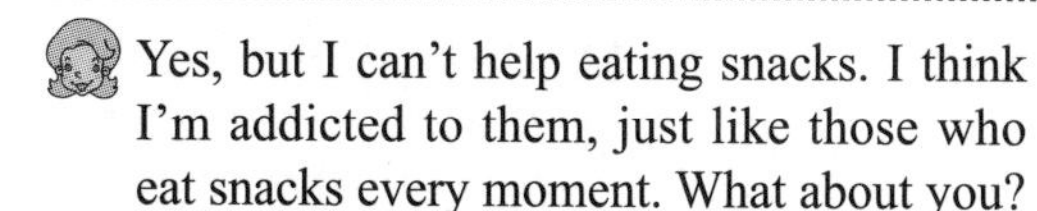

Yes, but I can't help eating snacks. I think I'm addicted to them, just like those who eat snacks every moment. What about you?

是的，但是我控制不了，我想我已经上瘾了，就像那些人一样，每时每刻都在吃零食。你呢?

I don't like snacks at all. I have my meals on time and never eat snacks between meals. Most snacks are unhealthy, like crisps or ice cream.

我一点儿也不喜欢吃零食。我平时准时吃正餐，从不吃零食。大部分零食都是不健康的，比如薯片和冰淇淋。

That's true.

那没错。

开胃词组

run out of 用完

a pile of 一堆

be crazy about 对……疯狂, 痴迷

junk food 垃圾食物

be addict to 热爱……；沉溺于……

鲜美单词

supermarket n. 超级市场

snack n. 快餐，点心；加餐

spoil v. 损坏，糟蹋

appetite n. 胃口，食欲

addict v. 使沉溺，使上瘾

crisp n. 薯片

Conversation 2

I feel like a bit hungry.	我有点饿了。
But it's not time for dinner yet.	但是还不到吃饭时间呢。
Do we have anything to eat?	我们有吃的东西吗?
How about some snacks?	吃点儿零食怎么样?
I think snacks contain a lot of fat.	我觉得零食包含许多脂肪。
Well, I recommend some special snacks, which are made from veggies.	我推荐一些零食给你，它们是蔬菜做的。
Are they tasty?	好吃吗?
They're in different flavours, spicy or salty. Some of them are preserved or dried.	它们有不同的口味，辣的，咸的。一些是腌制的或干货。
OK. I'll give it a whirl.	好啊，我尝尝。

开胃词组

it's time for 该是……的时候

be made from 由……做成的

give it a whirl 尝一尝，试一试

鲜美单词

hungry adj. 饥饿的

dinner n. 正餐，晚餐

contain v. 包含，含有

recommend v. 推荐，介绍

veggies n. 蔬菜

flavour *n.* 风味，味道

preserve *v.* 保存，腌制

不饿的时候吃零食，很容易发胖

Sometimes a snack quiets your growling stomach. But sometimes you just feel like munching, even though you're not really hungry. French researchers recently found that this second type of snack attack is harder on your waistline.

有时，吃点零食能让你的肚子不再咕咕叫。但有些时候，你只是想要吃点东西，尽管你并不饿。法国研究人员最近发现，第二种情况很容易让你发胖。

That's because when you eat without being hungry, blood sugar and insulin work differently than when you are hungry.

这是因为当你不饿的状态下吃东西时，血糖和胰岛素的变化与你饥饿的时候是不同的。

First, let's see what happens when you're hungry. Low blood sugar can trigger that hungry feeling. When you eat, your blood sugar rises, satisfying your hunger and prompting your pancreas to secrete insulin. Insulin's job is to transport sugar from the bloodstream into your body cells. Hours later, after insulin has moved enough sugar into the cells and leveled off, your blood sugar dips and you feel hungry again.

首先，让我们看看当你饥饿时的情况。这时，低血糖会让你产生饥饿的感觉。在你吃东西的时候，你的血糖会上升，消除你的饥饿感并促使胰脏分泌胰岛素。胰岛素的作用是把血液中的糖分运送到你体

细胞中。几小时后，当胰岛素把足够多的糖分送入细胞内并趋于稳定时，你的血糖量就会降下来，然后你又会开始感到饥饿。

But let’s say you snack before you feel hungry again. What happens to your blood sugar then? The French study found that if people snacked without being hungry, their blood sugar stayed the same as if they hadn’t snacked at all. Why? Since the pancreas is still secreting insulin to transport blood sugar from the previous meal, it’s easily stimulated to secrete a little more insulin.

让我们再来说一说当你不饿时吃零食的情况吧。你的血糖会怎样变化呢？法国研究表明，如果人们在不饿的情况下吃零食，他们的血糖会保持不变，就像自己从没吃过零食一样。为什么会这样呢？这是由于胰脏仍在分泌胰岛素来输送上一顿饭中的血糖，这时候吃零食很容易刺激胰脏分泌更多的胰岛素。

People’s insulin levels rose, but their blood sugar stayed the same. So, they got hungry at about their usual time, and ate just as much as when they didn’t snack.

人体的胰岛素含量会上升，但是血糖浓度却保持不变。因此，人们仍然会在平常吃饭的时间感到饥饿，饭量与他们不吃零食时一样多。

If you’re like the folks in the study, snacking when you’re hungry is okay. But when you’re not hungry, snacking probably won’t satisfy you or make you eat less later. Instead, it’ll prompt you to loosen that belt a bit.

如果你和那些人一样，在饿的时候才吃零食，是可以的。但如果你不饿，吃零食可能不会给你饱足感或者让你随后少吃点。相反，它会让你敞开胃口，吃得更多。

UNIT 05

Balanced Nutrition
营养均衡

舌尖美食词汇

nutrition 营养	**calorie** 卡路里
protein 蛋白质	**carbohydrate** 碳水化合物
vitamin 维生素	**green vegetable** 绿色蔬菜
starchy food 淀粉食物	**dairy product** 乳制品
cereal 谷类食物	**minerals** 矿物质
calcium 钙	**ferrum** 铁
additive 添加剂	**quota** 定量

舌尖美食句

1. It's important to keep a balance in nutrition.	营养均衡很重要。
2. Calories come from three main sources: carbohydrates, proteins and fats.	卡路里有3种主要来源：碳水化合物、蛋白质和脂肪。
3. A healthy, balanced diet is integral for weight control.	健康的营养均衡的饮食是减肥的最基本要求。

4.	You should eat more vegetables and fruits to get enough vitamins.	你应该多吃点蔬菜水果，摄取足量的维生素。
5.	Do you eat a lot of meat everyday?	你每天都吃很多肉吗？
6.	We should have enough starchy food to intake fiber, vitamins and minerals.	我们应该食用足够的淀粉食物以摄取纤维、维生素和矿物质。
7.	Foods that may boost brain function and development include dairy products, fruits, leafy green vegetables, fish, nuts and liver organ.	能促进人脑的机能和生长的食物包括乳制品、水果、绿叶蔬菜、鱼、果仁和动物肝脏。
8.	She is not fussy about her food.	她不挑食。
9.	Water is also vital nutrients of the body.	水也是人体的重要营养成分。
10.	Eating cereals and fruit will give you plenty of fiber in your diet.	吃谷类食物和水果能多摄取纤维。

舌尖聊美食

Conversation 1

	Oh, Anna. You said you're on a diet, but look at the food on your tray!	噢，安娜，你说你在减肥，瞧你餐盘里的东西。
	What's wrong with it?	怎么啦？
	You should keep a balance in nutrition.	你应该保持营养均衡。
	Yes, I did. Look, I got protein from the hamburger and fiber from the salad. It looks quite balanced to me.	对呀，我是这样做的啊。看，我从汉堡中摄取蛋白质，从沙拉中摄取纤维，看上去饮食挺均衡的。

You should watch your intake of fat, sugar and calories. Normally we need 2,000 calories per day. Let's take a look at your food: a cheeseburger has about 400 calories; a large soda has about 200 calories; those steaks have at least 800 calories.

你要注意脂肪和糖的摄入量。通常我们每天需要2000千卡的热量，看看你的食物：芝士汉堡大约400千卡，大杯苏打水约200千卡，牛排至少800千卡。

For this meal, I have got 1400 calories already. So I can only have 900 calories for the dinner.

这么看来，我这顿饭已经1400千卡了，晚饭只能吃900千卡。

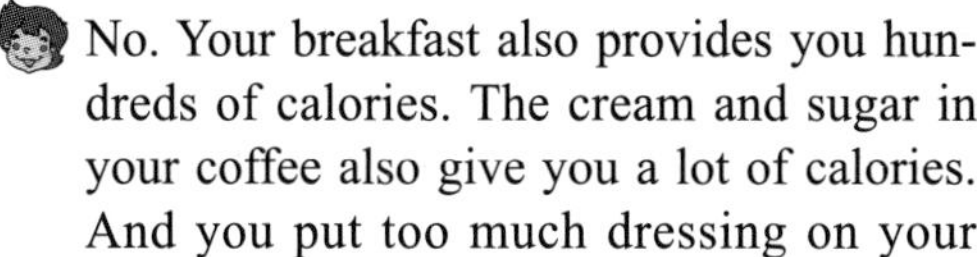
No. Your breakfast also provides you hundreds of calories. The cream and sugar in your coffee also give you a lot of calories. And you put too much dressing on your steaks. You not only watch the number of calories but also the amount of fat and sugar you have every day.

不对，早餐还提供给你好几百千卡的热量呢。你咖啡里的奶油和糖也提供许多热量，还有你的牛排放太多酱了。你每天不但要注意摄入卡路里的多少，还要注意脂肪和糖的摄入量。

How can I deal with this problem?

那我该怎么解决？

Well, remember to eat food that's high in fiber, and low in fat and sugar.

嗯，记住吃含高纤维、低脂肪的食物和糖。

I see. Thanks.

明白，谢谢。

开胃词组

on a diet 节食

keep a balance 保持平衡

at least 至少，最少

hundreds of 几百的，数百的

not only…but also… 不但……而且……

鲜美单词

tray *n.* 盘子，托盘
balance *n.* & *v.* 平衡，均衡
nutrition *n.* 营养
fiber *n.* 纤维，纤维物质
intake *n.* 摄取量，吸取量
normally *adv.* 通常地，一般地
steak *n.* 牛排，肉排
provide *v.* 提供，供给
dressing *n.* 调料，调味品
deal *v.* 应付，处理

Conversation 2

I think I may try to eat a balanced diet.	我觉得我应该吃营养均衡点的食物。
Me too. I have changed my diet recently, and I eat a lot healthier now.	我也是，我最近已经改变了饮食，吃得更健康了。
What do you eat?	你都吃些什么？
My diet consists mainly of fruits, veggies and chicken.	我的食物主要有水果、蔬菜和鸡肉。
That's it?	就这些？
Yeah. Fruits and veggies give me fiber and chicken meat gives me protein and some minerals.	是的，水果和蔬菜提供纤维，鸡肉提供蛋白质和一些矿物质。
No bread and no fried meat?	不吃面包和油炸的肉类吗？

Um, I would like to eat some bread; the whole wheat bread is better, but no fried meat which is rich in fat.	嗯，我吃一些面包，最好是全麦面包，但不吃油炸的肉，那含脂肪太多了。
So, what about the drink, coke or coffee?	那么，喝什么饮料，可乐或咖啡吗?
None of them, just water or soup. Coke has a lot of carbohydrate, and milk coffee contains fat, too.	都不要，只喝水或汤就可以。可乐碳水化合物含量太高，牛奶咖啡也含有脂肪。
That sounds a bit hard for me.	这对我来说太难了。
You should try it. A balanced diet is good for your health.	你应该试一试，营养均衡的饮食让你更健康。

开胃词组

consist of 由……组成

whole wheat 全麦

be rich in 含有丰富的……，富含

none of 都不

鲜美单词

mainly *adv.* 大部分地，主要地

chicken *n.* 鸡肉

protein *n.* 蛋白质

mineral *n.* 矿物，矿物质

meat *n.* 肉，肉类

carbohydrate *n.* 碳水化合物

coffee *n.* 咖啡

healthy *adj.* 健康的，健壮的

Food Pyramid食物金字塔

要保持身体健康，需要均衡的饮食。人体所需要的营养物质来自于每天的食物，科学家根据人体机能及能量消耗量，形成了一个食物金字塔，展示每天人体所需的食物种类及数量，其中，金字塔底部表示人体所需的食物量较大，而金字塔顶部表示人体所需的食物量较少：

1. Fat, Oils, and Sweets

脂肪，油，糖

2. Milk, Cheese and Yogurt Group (2 to 3 servings per day)

牛奶、奶酪、酸奶类（每天2到3份）

3. Meat, Poultry, Egg, Fish, Beans and Nuts Group (2 to 3 servings per day)

肉类、家禽肉、鸡蛋、鱼肉、豆类和坚果类（每天2到3份）

4. Vegetable Group (3to 5 servings per day)

蔬菜类（每天3到5份）

5. Fruit Group (2 to 4 servings per day)

水果类（每天2到4份）

6. Rice, Bread, Cereal and Pasta Group (6 to 11servings per day)

米饭、面包、麦片和面食类（每天6到11份）

Part 2

Buying Ingredients and Kitchenware

购买食材及厨具

Shopping for Vegetables 购买蔬菜

舌尖美食词汇

fresh leafy greens 绿色蔬菜	**tomato** 西红柿
ingredient 食材	**organic products** 有机产品
broccoli 花椰菜	**cucumber** 黄瓜
carrot 胡萝卜	**potato** 土豆
eggplant 茄子	**pea** 豌豆
parsley 香芹	**lotus roots** 莲藕

舌尖美食句

1. May I help you? 需要帮忙吗？

2. We have all kinds of fresh leafy greens. 我们有各种绿色蔬菜。

3. Would you like some tomatoes? 你想来点西红柿吗？

4. I'm looking for some ingredients for my recipes. 我正在给我的菜谱找一些食材。

5. These vegetables are new on the list. 这些蔬菜是刚上市的。

6. Do you have any organic products?	你们卖有机食物吗?
7. All of broccoli is sold out.	所有的花椰菜都卖完了。
8. What's the price of these carrots?	这些胡萝卜什么价格?
9. Can you pick out the rotten parts, please?	你能挑出那些腐烂的部分吗?
10. What else would you like?	你还要别的吗?
11. I wonder if you have any cucumbers.	我想知道你卖不卖黄瓜。
12. Those potatoes have sprouted.	那些土豆都发芽了。

Conversation 1

How can I help you?	需要帮忙吗?
I want to buy some vegetables.	我想买些蔬菜。
Well, we have all kinds of fresh leafy greens. What do you need?	嗯，我们有各种绿色蔬菜，你想要什么?
I am a vegetarian I eat starchy food every day.	我是一个素食者，我每天都吃些淀粉食物。
Um, these potatoes are new here and they're rich in starch.	这些土豆是新来的，它们是富含淀粉的食物。
I ate potatoes yesterday. Anything for a change?	我昨天刚吃了土豆，有其他的吗?

How about lotus roots? They're also a kind of starchy food.	这些莲藕怎么样？它们也是淀粉食物。
Excellent. I will take one pound. How much are these lettuces?	棒极了，我要一磅。这些生菜多少钱？
They're 3 dollars each pound.	每磅3美元。
Could you offer any discount?	可以打折吗？
Sorry, ma'am. These're the best selling organic products these days, not any discount.	对不起，夫人。这可是最近最畅销的有机食物，不打折的。
What if I bought more two pound of string beans, can I get a discount for them?	如果再买两磅青豆角，可不可以打折呢？
In that case, we give you 10 percent discount.	如果那样的话，给你打九折。
OK.	好的。

开胃词组

all kinds of 各种各样的

every day 每天，每一天

best selling 最畅销的

in that case 如果那样的话

鲜美单词

leafy *adj.* 多叶的，叶茂的

vegetarian *n.* 素食者

starchy *adj.* 淀粉含量高的

lotus *n.* 莲花

lettuce *n.* 莴苣，生菜

discount n. & v. 折扣；打折扣
organic adj. 有机的；绿色的
string n. 串，一串串；线，细线

Conversation 2

Good morning, ma'am. Welcome to Frank's Vegetables Shop.	早上好，夫人，欢迎来到弗兰克蔬菜店。
Good morning, I am looking for the ingredients of corn roll. Do you have any corn here?	早上好，我想买点玉米卷的原料。这里有玉米吗？
I'm sorry, all corns have been sold out.	对不起，所有玉米都卖光了。
What a shame!	真遗憾！
Would you like some cabbages? Cabbages taste good in fried.	你来点卷心菜吗？卷心菜炒着好吃。
Well, I prefer carrots fried with chicken. They're nice in salad, too. Anyway, I just can't get enough carrots to eat. I take two pound of these carrots.	嗯，我喜欢胡萝卜炒鸡肉。胡萝卜做沙拉也好吃。不管怎样，我吃胡萝卜就是吃不够。我买两磅这些胡萝卜。
The carrots are two dollars a pound. They're almost gone. How about giving you the rest of them for 3 dollars?	胡萝卜每磅2美元，这些都快卖完了，3美元全卖给你怎么样？
All right, you mean in total?	好的，你是说全部这些？
Yes, all of these only cost 3 dollars.	是的，这些都给你，共3美元。
OK. Here's the money.	好的，给你钱。

Here is your change. Thank you for coming and have a nice day.

找你零钱。谢谢您的光临，祝您过得愉快！

Thanks, you too.

谢谢，你也是。

开胃词组

corn roll 玉米卷

sell out 卖完，售罄

what a shame 真遗憾

the rest of 剩余下的

in total 总计，一共

鲜美单词

ingredient n. 原料，材料

cabbage n. 卷心菜，圆白菜

enough n. 充分，足够

carrot n. 胡萝卜

salad n. 沙拉

anyway adv. 不管怎样，无论哪种方式

total n. 总数，总计

change v. 改变，变换

舌尖美食文化

How to Shop for Fresh Vegetables

如何购买新鲜蔬菜

Many people choose to take a risk to buy some vegetables which may be rotten. Getting home to discover that your potatoes are rotten, or the ears of corn are moldy, it is aggravating and a waste of good money. To avoid that aggravation and waste, we offer some rules for buying fresh vegetables.

许多人不喜欢去挑选新鲜的蔬菜而是冒冒失失地买了菜，后来才发现买了腐烂的蔬菜。回到家，人们发现自己买的土豆已经坏了，买的玉米已经发霉了，这真是一件既恼人又费钱的事。为了避免这些悲剧和浪费，这里提供一些关于购买蔬菜的温馨提示。

1. Examine each vegetable individually. Sometimes there will be a single rotten piece mixed in with a good group. Be selective.

单个检查每个要购买的蔬菜，往往一个腐烂的菜总是混在好的里面，所以要精挑细选。

2. Look for bright color. Darkened coloring and browning is a sign of age in vegetables and means that your shelf life at home will be shortened.

寻找那些颜色亮泽的蔬菜。深色或棕色的，一般放了很久，意味着拿回家就储藏不了多久。

3. Check for firmness and crispness. Vegetables should be firm when you gently squeeze them. Avoid wilted looking greens and celery.

查看是否饱满或易碎的。蔬菜捏上去是饱满的，就说明它是新鲜的。不要购买发蔫的蔬菜或香芹。

4. Examine ears of corn by pulling off the husks in the store. Corn husks should be fresh and succulent with good green color. Avoid ears with under-developed kernels, and old ears with very large kernels.

在商店购买玉米时，拔下它的穗查看一下。新鲜的玉米外壳应该是鲜嫩多汁的、饱满绿色的。不要买穗没成熟的玉米，也不要买穗老粒大的玉米。

5. Choose young mushrooms that are small to medium size. Caps should be mostly closed around the stem, and white or creamy, or

uniformly brown if a brown type.

选购大小适中的鲜嫩蘑菇。选择那种菇伞还没打开的，白色的或奶油色的，或整体棕色的蘑菇。

6. Avoid potatoes that have decay, blemishes, green, or have too many cuts in the skin. While decay can be cut from potatoes, you want to avoid excessive waste.

不要选购那些腐烂的、有斑点的、发绿的或表皮有切口的土豆。土豆的腐烂部分被去除后，所剩无几，那真是浪费。

读书笔记

Buying Meat 挑选肉类

舌尖美食词汇

fresh meat 鲜肉	**sirloin** 牛里脊肉
ground beef 牛肉馅	**round** 牛骨肉
brisket 牛胸肉	**chicken wing** 鸡翅
chicken drumstick 鸡腿	**chicken breasts** 鸡胸
lamb 羊肉	**ribs** 肋排
filet 去骨肉	**pork chop** 猪排

舌尖美食句

1. I will go shopping for some fresh meat.	我要去买些鲜肉。
2. Do you have any sirloin?	你们有牛里脊肉吗?
3. I will take a pound of ground beef.	我要一磅牛肉馅。
4. Would you like some chicken wings?	你想来点鸡翅吗?
5. Could you take the skin off for me?	你能帮我去皮吗?
6. I want two pieces of chicken breasts.	我要两块鸡胸肉。

7. These beef is just on sale this morning.	这些牛肉今天早上刚上市。
8. How much do you need?	你需要多少？
9. That's enough for me.	对我来说足够了。
10. What part of the lamb is this?	这是什么部位的羊肉？
11. How about those ribs?	那些肋排怎样？
12. Is anything else I can get for you?	你还需要别的吗？

舌尖聊美食

Conversation 1

Can I help you?	你好，想买点什么？
I'd like half a pound of ground beef, please.	我要半磅牛肉馅。
OK. Our ground beef is quite lean.	好的，我们这儿的牛肉很瘦。
I would like half a dozen pork chops and two pounds of boneless chicken breasts. Do you have them?	我还要半打猪排，2磅无骨鸡胸，你这里有吗？
No, all chicken breasts have been sold out, but we have some nice chicken drumsticks.	没有，所有的鸡胸都卖完了，但是有很好的鸡腿。
No, that's not what I want. I'll take this smoked ham you have here.	那不是我想要的，我买这块熏火腿。
Okay, is there anything else?	好的，还要些别的吗？

Is this so called salami and bologna you have here?	这就是意大利咸味腊肠和大腊肠吗？
Yes! It's very fine meat! Made it myself...	是的，很好的肉，我们自己做的……
Sounds good. Okay, that's it.	听起来不错。就买那个了。
Wait! We have T-bone, rib eye, and sirloin steaks. They are very fresh! They just came here today.	等一下，这里有牛排，牛里脊肉和牛腰肉，都很新鲜，今天刚运来的。
Mmm...No, that's fine. I think that's all for today.	嗯，不，不要了，今天就买这些吧。
Okay. That will be sixty-five dollars and fifty cents.	好的，一共是65美元50美分。

开胃词组

a pound of 一磅……

half a dozen 半打

T-bone 丁字牛排

rib eye 牛里脊肉

smoked ham 熏火腿

鲜美单词

lean adj. 瘦的

pork n. 猪肉

boneless adj. 去骨的

breast n. 胸部

drumstick n. 鼓槌；鸡腿

salami n. 意大利腊肠

rib *n.* 肋骨

bologna *n.* 大腊肠

sirloin *n.* 牛里脊肉

steak *n.* 牛排

Conversation 2

Can I help you?	需要买什么吗?
How much are those ribs?	肋排怎么卖?
5 dollars a pound. How much do you need?	一磅5美元，你需要多少?
I will take three pounds. I guess it should be enough for four people.	我要3磅，我猜应该够四个人的。
You must be an experienced chef! Is there anything else?	你肯定是一个经验丰富的厨师！还要别的吗?
Yeah, I'm thinking about some meat for hot pot.	对，我在考虑买一些下火锅的肉。
We have some fresh and tender cutlets, good for hot pot.	我们有又鲜又嫩的羊肉片，正是下火锅的好料。
Good idea. And now I got ribs and cutlets which are both a bit heavy. I like to have something light, like chicken salad.	好主意。我买了肋排和羊肉片，全是口味比较重的。我想做点清淡的食物，比如鸡肉沙拉。
Well, here is some fresh chicken meat.	嗯，这里有一些新鲜的鸡肉。
What part of the chicken is this?	这是什么部位的鸡肉?

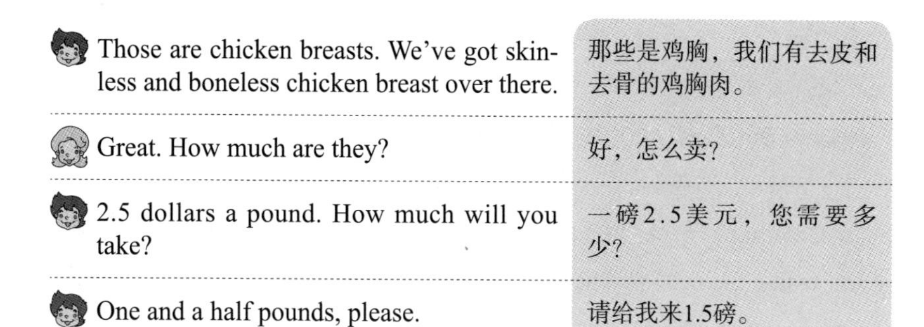

Those are chicken breasts. We've got skinless and boneless chicken breast over there.	那些是鸡胸，我们有去皮和去骨的鸡胸肉。
Great. How much are they?	好，怎么卖?
2.5 dollars a pound. How much will you take?	一磅2.5美元，您需要多少?
One and a half pounds, please.	请给我来1.5磅。

开胃词组

hot pot 火锅

good idea 好主意

chicken salad 鸡肉沙拉

over there 在那边

鲜美单词

chef *n.* 厨师

tender *n.* 软的，嫩的

cutlet *n.* 肉片

heavy *adj.* 重的，浓烈的

salad *n.* 色拉

skinless *adj.* 无皮的

boneless *adj.* 去骨的

How to Buy Beef? 如何选购牛肉?

When buying beef, consider its grading and cut. Also, buy beef with good marbling (flecks of fat) — the more marbling there is, the more moist and tender the beef will be. Beef is graded according to the animal's age, the amount of marbling in the cut, and the beef's color and texture.

根据牛肉的等级和切割来选购牛肉。优质的牛肉一般带有良好的纹理（脂肪斑纹）——纹理越多的牛肉，越鲜嫩多汁。牛肉等级是按牛的年龄、纹理数、肉色和质地来划分的。

Most often, you'll find three grades of beef when shopping:

当你购物时你会发现牛肉一般分为三个等级:

Prime: The highest grade and the most expensive. In general, the most tender and flavorful beef falls under this category. Most beef from this grade end up at hotels and fancy restaurants. You can also find it at specialty butcher shops.

上等牛肉：最高等级和最贵的牛肉。一般而言，这个等级的牛肉最鲜嫩、最美味。这个等级的牛肉一般供应酒店和豪华餐厅。在一些专业肉铺你也能买到。

Choice: The second tier of beef grading, leaner than prime. This is what you'll find most often at the supermarket.

优选等级牛肉：这是第二等级牛肉，比第一个等级更精瘦。这个等级的牛肉大多数超市均有售。

Select: Best for stewing and braising. You'll find this in supermarkets.

精选等级牛肉：炖牛肉和烧牛肉的好材料。超市均有售。

The more tender cuts of beef include steaks such as porterhouse, sirloin, shell, New York strip, Delmonico and filet mignon-as well as roasts such as rib, rib eye and tenderloin Tender meats are usually cooked by the dry heat methods of roasting, broiling, grilling and sautéing.

最鲜嫩的牛肉包括牛排，比如腰肉牛排、牛里脊肉、去皮牛肉、纽约牛排、德蒙尼克牛排和牛肉卷，还包括烤牛肉，比如肋排、肋眼牛排、里脊肉。鲜嫩牛肉一般用火炙烤、焙烧、烧烤和爆炒。

Look beyond ratings to judge meat. Meat should look bright red, never dull or gray. Excess juice in the package may indicate that the meat has been previously frozen and thawed - don't purchase it.

除了看等级，牛肉还可通过观察外表来判断其质量优劣。好的牛肉看起来是鲜红色的，而非暗色或灰色的。肉里渗出水的，表明牛肉经过冷冻或者解冻，谨慎购买。

读书笔记

UNIT 03

Buying fruits 买水果

舌尖美食词汇

seasonal fruit 时令水果	**orange** 橙子
peach 桃子	**grape** 葡萄
blueberry 蓝莓	**watermelon** 西瓜
origin 产地	**apple** 苹果
pear 梨	**coconut** 椰子
mango 芒果	

舌尖美食句

1. Do you have any seasonal fruits?	你这里有时令水果吗？
2. Those oranges look a bit saggy.	那些橙子看起来松垮垮的。
3. Can I have half kilo of strawberries?	我可以来半公斤草莓吗？
4. Are peaches in season right now?	桃子是现在的当季水果吗？
5. These grapes taste sweet. Would you like to try some?	这些葡萄很甜，要不要品尝一下？

6. That's exactly two pounds.	正好两磅。
7. You can get a bag of blueberries for two bucks.	一袋蓝莓价钱是两美元。
8. What's the origin of these bananas?	这些香蕉的产地是哪里？
9. Can I try some?	我可以尝一下吗？
10. All the strawberries just have been sold out.	所有的草莓都卖完了。
11. These watermelons are ripe enough and juicy.	这些西瓜熟透了，很有水分。
12. Can you give me a better price?	能再便宜点吗？

舌尖聊美食

Conversation 1

I wanna buy some fruits in season.	我要买些时令水果。
The grape and watermelon are seasonal fruits now.	葡萄和西瓜是当前的时令水果。
Can I sample a grape?	我能尝一个葡萄吗？
Sure.	当然可以。
It tastes juicy and sweet. What's the origin of them?	尝起来还挺甜且多汁的。它们的产地是哪里的？
Well, these are from French, the home of grapes.	嗯，这些葡萄来自法国，葡萄之乡。
How much are they?	怎么卖？

2 dollars a pound.	每磅2美元。
I will take 3 pounds. And wrap it, please.	我要3磅，请打包好。
No problem.	没问题。
Here is the money and keep the change.	给你钱，不用找零了。
Thank you.	谢谢!
You're welcome.	不客气。

开胃词组

in season 时令的，应时的

the origin of ……的产地；……的起源

keep the change 不用找零钱了

鲜美单词

season *n.* 季，季节

grape *n.* 葡萄

watermelon *n.* 西瓜

juicy *adj.* 多汁的

origin *n.* 起源，来源；产地

wrap *v.* 包，包装

change *n.* 零钱

Conversation 2

Hello, Ms. Gordon. How are you today?	你好，戈登女士，你怎么样?

Not too bad, Henry. How's your business?	还好，亨利。生意怎么样？
Pretty good. All the strawberries just have been sold out.	不错。所有的草莓都卖完了。
Good news. I come to buy some blueberries for making fruit juice.	真是好消息呀。我买点蓝莓做果汁。
These blueberries are suitable for fruit juice.	这些蓝莓做果汁不错的。
They look the same as what I bought last time. They are mushy and not very sweet.	它们看起来跟上次买的一样，软绵绵的而且不甜。
Well, I got some choice blueberries which are from Spain, but they're much more expensive.	那么，我还有一些上等的蓝莓，来自西班牙的，但是更贵些。
Can I try some?	我可以尝一下吗？
Yeah, sure.	当然可以。
Um… They taste nice and juicy. I will take two boxes.	嗯，吃起来甜美多汁。我要两盒。
That costs you ten bucks. That's all?	那得要十美元。还要什么吗？
I'd like to get a couple of kiwi fruits.	我想再买点猕猴桃。
These products are promoted. One box is two bucks.	这些猕猴桃正在搞促销，每盒只收两美元。
Really? I take three boxes for that. I think that's all for today. How much do I own you?	真的吗？我来三盒。我想就这些了。一共给你多少钱？
Let me see. Two boxes of blueberries and three boxes of kiwis… That will be sixteen bucks.	我看看，两盒蓝莓和三盒猕猴桃……一共十六美元。

Here you go. Thanks a lot, Henry. I'll see you.

给你钱。谢谢你，亨利。再见！

Good day to you, Ms. Gordon.

祝你愉快，戈登女士。

开胃词组

good news 好消息，喜讯

fruit juice 水果汁

be suitable for 适合……

look the same 看起来一样

here you go 给你

鲜美单词

business *n.* 商业，生意

strawberry *n.* 草莓

juice *n.* 果汁

suitable *adj.* 合适的，适当的

mushy *adj.* 软塌塌的

Spain *adj.* 西班牙的；西班牙人

expensive *adj.* 昂贵的

buck *n.* 美元

promote *v.* 促销，推广

Guides to Buying Fresh Fruits 新鲜水果购物指南

We need to eat high quality, nutritious fresh produce. At the same time, we need to spend less money, reduce our grocery costs and reduce wasteful spending. So, how can we select high quality fruit in the market, make sure it is ripe, and then keep it fresh at home? There are a few basic rules:

我们希望花最少的钱就能吃到高品质的、有营养的新鲜食物。那么，我们怎样在市场里挑选高品质的水果，确认它是否成熟了并拿回家保鲜呢？以下有几个小提示：

1. Buy fruit when it is in season where you live. Produce can be marked “in season now” when it’s imported from another country or location where it is in season, but that doesn’t mean it will be ripe and flavorful by the time it arrives in your home.

购买时令的水果。水果上有的标注“时令水果”，也许那是从国外进口或别的地方过来的，并不意味着它们被你买回家时已经是成熟的、可口的。

2. Never seal fruits in an air tight bag. Fruits needs to “breathe”, with air circulation, and air tight containers speed up the decaying process.

不要把水果放袋子里裹得严严实实的，水果也需要靠空气循环来“呼吸”。在密封的容器里，水果更容易变质。

3. Some fruits emit ethylene, a natural gas that speeds ripening. Other fruits may be extremely sensitive to ethylene, and will decay within days if stored together with ethylene-producing fruits.

一些水果会散发乙烯，一种自然催熟的气体。其他的水果可能对这种气体有反应，如果把它们堆放一起，容易变质。

Choosing Sauces and Dressings 选购酱料

舌尖美食词汇

flavor 口味	**sesame paste** 芝麻酱
salad dressing 沙拉酱	**blue cheese** 蓝纹乳酪酱
ketchup 番茄酱	**sweet bean sauce** 甜面酱
chili sauce 辣椒酱	**Caesar salad** 凯撒沙拉酱
cooking wine 料酒	**oyster oil** 蚝油

舌尖美食句

1.	What's your favorite flavor?	你喜欢什么口味的?
2.	Do you have sesame paste?	你们有芝麻酱吗?
3.	Would you like to try our new salad dressing?	你想试试我们新出的沙拉酱吗?
4.	I'd like to choose something light.	我想选清淡些的。
5.	The ketchup is on the second layer of the shelf in aisle 3.	番茄酱在第三过道货架的第二层。

6.	There are thirty different flavors of dressings here.	这里有三十种不同口味的调味品。
7.	The bottle cap has been loosed. Can I change another one?	这个瓶盖已经松开了。我可以换一瓶吗?
8.	These items are on sale. You buy one and get another one free.	这些货品在减价出售。买一赠一。
9.	This blue cheese has exceeded the expired date.	这个蓝纹乳酪已经过期了。
10.	She was addicted to sweet bean sauce.	她喜欢甜面酱。
11.	There's no price tag on this chili sauce. Could you check the price for me?	这瓶辣椒酱没价格标签，你能给我看一下价格吗?

舌尖聊美食

Conversation 1

Excuse me. I can't find the dressing in this supermarket.	打扰一下，在这个超市，我怎么找不到调味酱在哪里?
The dressing is on the shelf of condiment section. There is a testing stand there. I'd be glad to show you the way.	调味酱是在酱料区。那里有个样品摊位。我很乐意带你去。
Thank you very much.	谢谢。
Here they are. All flavors of dressing are available here. There must be one dressing pleasing you. And you can also have a taste at the testing stand.	这里就是。各种口味的酱料都在这里，肯定有一种让你满意。你也可以在样品台这里品尝一下。

Wow. This is amazing. Let me see, barbecue sauce, ketchup, mustard relish, Guacamole, steak sauce, sesame paste... there must be thousands of dressing here.	哇!太神奇了。我看看，烤肉酱、番茄酱、芥末、鳄梨酱、牛排酱、芝麻酱…这里肯定得有上千种酱料。
Yes, we have three shelves of different dressing. Would you like to try our new salad dressing? The creamy one is Caesar, and the red one is raspberry and walnut.	没错，我们三个货架摆放的全是不同品种的调味品。你想尝试一下新出的沙拉酱吗？这个含奶油的是凯撒沙拉酱，红色的是覆盆子胡桃酱。
I'm not a big fan of creamy dressing. Let me try the raspberry one...Mmm...it tastes kind of fruity.	我不太喜欢奶油酱，我来试试覆盆子。嗯……有水果味。
That's one of our "light choices". There are about ten other flavors.	这种口味很清淡，我们还有其他十种口味的。
I would like something sweet, like honey mustard.	我喜欢甜的，例如蜂蜜芥末酱。
Oh, that must be in the third aisle on the top shelf. Let's go and check it out.	哦，那个肯定是在第三过道的货架顶层。我们过去看看。
OK.	好的。

开胃词组

show the way 带路，引路

have a taste 尝一尝，品尝

a big fan of ……的粉丝；喜欢

check out 看看，查看

鲜美单词

supermarket *n.* 超级市场
dressing *n.* 调味品
condiment *n.* 调味品，酱料，佐料
stand *n.* 台，架子
flavor *n.* 味，味道
barbecue *n.* 烤肉，烧烤
ketchup *n.* 番茄酱
mustard *n.* 芥末，芥菜
guacamole *n.* 鳄梨酱
paste *n.* 糊状物；酱
raspberry *n.* 覆盆子，树莓
walnut *n.* 胡桃
aisle *n.* 过道，通道

Conversation 2

Wow, there're so many dressings to choose from here. What do you think?

哇，这里有很多调味品可供选择。你觉得应该怎么选？

Just go ahead with the basic one. As I know, Molly loves sweet and rich dressings. Maple thyme would be the right one for her.

就选常见的口味吧。我知道莫利喜欢甜的、浓的酱料，枫糖香草酱很适合她。

For me it's a toss-up between citrus vinaigrette and Italian I like my dressing sour and light. Well, I think citrus vinaigrette would be better. What do you like?

我个人的话柑橘醋酱和意大利酱两者选其一。我喜欢酸的和清淡的。嗯，我觉得柑橘醋酱好些吧。你喜欢什么酱？

I've got a bottle of cream sauce at home.

家里已经有一罐奶油沙拉酱了。

That should be enough. Let's pick other stuff and get out of here.

应该够了吧，我们再买其他的东西就走吧。

开胃词组

go ahead with 继续（进行）

as I know 据我所知

a toss-up between…and 两者之间选择；要么……要么……

get out of 离开……

鲜美单词

choose v. 选择，决定

basic adj. 基本的，根本的

maple n. 枫树

thyme n. 百里香

toss-up n. 两者择一

citrus n. 柑橘

舌尖美食文化

Low-calorie Salad Dressing 低热量沙拉酱

A salad is a dish of vegetables, legumes, eggs, pasta, meat, seafood or any mixture that is tossed together, and seasoned with spices, herbs or sauces. Salad dressings may be kitchen made or commercially manufactured. Salads are extremely popular among weight watchers and fitness freaks. Popular salad dressings include vinegar, olive oil, mayonnaise, lemon juice, cheese, yogurt, etc. Most of these dressings have high calories and may hinder our weight loss efforts. Although the basic ingredients of most salads consist of fresh, unprocessed food items such as vegetables and sprouts, the dressing used may be fatty and full of calories.

沙拉是由蔬菜、豆类、鸡蛋、意大利面、肉、海鲜或其他食物原料混合，再配上佐料、香料或酱油制作而成的。而沙拉酱可以家庭自制也可以工厂加工制作。沙拉受到瘦身减肥人士的推崇。受欢迎的沙拉调料有醋、橄榄油、蛋黄、柠檬汁、奶酪、酸奶等等。这些沙拉酱大多含高热量，或许会抵消减肥效果。尽管常见的沙拉一般由新鲜食物原料制作，如蔬菜、芽苗，但是沙拉酱却含有脂肪和高热量。

Nowadays, many different types of commercial low-calorie versions of traditional salad dressings are available. These ready-made dressings come in a wide variety of flavors and you'll have a hard time deciding which ones NOT to include in your salad bar. Apart from these, you can put together some simple and easily available ingredients in your own kitchen and whip up your very own low-calorie dressing. Here is one recipe that you can try out.

如今，许多低热量的沙拉酱到处都可以买到。这些沙拉酱口味不同，让人难以择其一。除了购买沙拉酱，你也可以在家自制属于自己的低热量沙拉酱，下面是一个沙拉酱食谱。

Creamy Balsamic Dressing 奶油香醋酱

Ingredients 原料：

¼ cup, Balsamic vinegar 四分之一杯的香醋

1 tablespoon, light soy sauce 1汤匙的酱油

3 tablespoons, fat-free yogurt 3汤匙脱脂酸奶

1½ tablespoons, Dijon mustard 1汤匙半的法式芥末

1½ tablespoons, honey　1汤匙半的蜂蜜

½ tablespoon, olive oil　半汤匙的橄榄油

Method: Mix all the ingredients well and store in the refrigerator.

制作方法：所有原料搅拌一起，放入冰柜储存。

Calories per Tablespoon: 31

每汤匙的热量：31

读书笔记

Buying Baking Ingredients 购买烘焙材料

舌尖美食词汇

baker 面包师	**donut** 甜面圈
pumpkin pie 南瓜派	**waffles** 华夫饼
vanilla extract 香草粉	**pumpkin powder** 南瓜粉
flour 面粉	**soda** 苏打（粉）
baking mix 套装烘焙粉	

舌尖美食句

1.	You used to be a baker, right?	你以前是一名烘焙师，对吗?
2.	Do you know how to bake donuts?	你会做甜面圈吗?
3.	May I have a pack of pumpkin powder?	可以给我来一包南瓜粉吗?
4.	I'm going to buy some ingredients for making biscuits.	我要去买点做饼干的材料。
5.	This is leftover vanilla extract we got here.	这是我们这里所剩的最后一点香草粉了。

6. I need some chocolate morsels to make chocolate chips.	我需要一些巧克力碎屑来做巧克力脆饼。
7. What kind of flour do you have?	你这里有哪种面粉?
8. Where can I shop for baking ingredients in Beijing?	在北京，哪里卖烘焙材料?
9. This all-in-one baking mix is great for making pancakes, waffles, biscuits, and many more American-style sweet and savory treats.	这个套装烘焙粉是制作薄煎饼、华夫饼、饼干等美国风味美食的最佳原料。
10. I can't find the almond extract for the cherry pie.	我找不到做樱桃派的杏仁粉。
11. The pumpkin powder is arrayed on the second shelf.	南瓜粉摆在第二个货架上。

舌尖聊美食

Conversation 1

 Are you learning how to make desserts?	你在学做点心吗?
 Yeah. I can make banana pudding, biscuits, and chocolate cheesecake. Last week we made peach pies. This week we're going to do brownies and cherry pies. OK. I've got the cocoa powder and vanilla extract, but I can't find the almond extract for the cherry pie.	是的，我会做香蕉布丁、饼干和巧克力奶酪蛋糕了。上周，我们做了桃子派。这周我们要做布朗尼和樱桃派。好，我找到了可可粉和香草粉，但是我找不到樱桃派用的杏仁粉。
 Look! It's right under your nose, next to the corn syrup.	看啊，它就在你跟前，玉米糖浆旁边。

Yes! I think I've got all the ingredients for the recipe. Do I miss anything?	对，我想食谱的所有材料都齐了。我没忘了什么吧？
Uh, I think we need more eggs. You have used the leftover eggs to make breakfast this morning.	嗯，我觉得我们得买点鸡蛋。今天早上你把剩下的鸡蛋用来做早餐了。
That's right. We need a dozen of eggs, and a pack of sugar, I think.	没错，我们再买一打鸡蛋，还有一袋糖。
OK. I'm looking forward to enjoying the cakes and pies now.	好啦。现在，我非常期待分享你做的蛋糕和派。
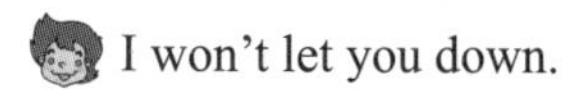 I won't let you down.	我不会让你失望的。

开胃词组

right under your nose 正好在你面前

next to 紧邻，挨着

make breakfast 做早餐

a dozen of 一打……

look forward to 盼望，期望

let sb. down 让某人失望

鲜美单词

dessert *n.* 餐后甜食，甜点

pudding *n.* 布丁

biscuit *n.* 饼干

cheesecake *n.* 奶酪蛋糕

brownie *n.* 布朗尼（美式糕点）

extract *n.* 提取物，浓缩物

almond *n.* 杏仁

syrup *n.* 糖浆，糖汁

ingredient *n.* 原料，材料

recipe *n.* 食谱，处方

leftover *adj.* 剩余的，用剩的

enjoy *v.* 享受，喜爱

Conversation 2

Hi, Lucy. Why do you have a lot of stuffs in your handcart?	你好，露西。你怎么拿那么多东西放小推车里。
Well, I will make some pumpkin pies and chocolate chips biscuits. These are the ingredients for my baking.	嗯，我要做南瓜派和巧克力脆饼。这些是我买的材料。
What have you got?	你都买了什么？
Well. For making pumpkin pies, I have got ground cinnamon, ground ginger, ground cloves and pumpkin powder. For the chocolate chips biscuits, I have bought some all-purpose flour, vanilla extract, chocolate morsels and baking soda.	嗯，做南瓜派，我买了肉桂粉、姜粉、丁香粉和南瓜粉。做巧克力脆饼，我买了通用面粉、香草粉、巧克力碎屑和小苏打。
That's a lot of materials. Baking job doesn't sound easy.	那可是很多材料。看来烘焙可不是简单的事情。
Yes, that's right. In fact, these are only a part of ingredients. I have the rest at home already, like butter, sugar…	没错，实际上，这只是其中的一部分材料。我在家还有其他的，如黄油、糖等等。
You must be very good at baking. How long have you been doing it?	你肯定很擅长烘焙。你学多久啦？
Not really. I'm just a beginner.	没有。我只是一个初学者。
Well, may I join you?	哦，我可以跟你一起做吗？

Sure. Come over tomorrow afternoon. 当然可以。明天下午过来吧。

开胃词组

pumpkin pie 南瓜派

ground cinnamon 肉桂粉

all-purpose flour 通用面粉

vanilla extract 香草粉

in fact 实际上，事实上

be good at 擅长

come over 过来，顺便来访

鲜美单词

handcart *n.* 手推车

ground *adj.* 碎的，粉末状的

cinnamon *n.* 肉桂

ginger *n.* 姜，生姜

clove *n.* 丁香

morsel *n.* 小块，碎屑

beginner *n.* 新手，初学者

舌尖美食文化

What's baking mix? 什么是烘焙混合套装

A baking mix is a mix of ingredients to which liquid, and sometimes oil, eggs or other ingredients are added to produce baked goods like muffins, biscuits, cakes and brownies. The earliest ones were made during the Industrial Revolution and proved helpful to those who lacked the time to stay home and carefully prepare food. These were commonly recipes for puddings or gelatin.

烘烤混合粉是一种烘焙混合套装原料，只需往里添加水、油、鸡蛋或其他原料就可以制作松饼、饼干、蛋糕或布朗尼。

最早的烘烤混合粉出现在工业革命时代，是为了帮助那些没时间在家精心准备食物的人。混合粉常用于制造布丁和果冻。

Biscuit and muffin baking mixes were soon offered, and were developed almost simultaneously. The big names in the US for these early mix forms were brands that are still familiar, like Jiffy and Bisquick. Betty Crocker offered the first mix varieties for cake in the 1920s.

饼干和松饼烘焙混合粉很快被研发出来，并广受欢迎。在美国，一些最早的烘焙混合粉品牌仍为人们所熟悉，比如Jiffy牌、Bisquick牌等。Betty Crocker牌蛋糕混合粉系列早在20世纪20年代就被研制出来。

In addition to saving time, many cooks, primarily women, preferred the predictability of the baking mix. With properly measured ingredients, the likelihood of turning out nice looking baked goods could be a big help. Early mixes often asked for the addition of numerous ingredients, but soon, many mixes came with powdered eggs, rendering the separating, cracking or beating of eggs unnecessary. You'll still find some varieties that require quite a bit of additional work. In fact, some are only slightly easier than the dry ingredients you'd mix on your own.

除了省时之外，许多烹饪师，甚至家庭主妇们，最喜欢这种混合粉的可靠性。这种混合粉经过精确测量，制作出来的东西非常可靠实用。刚开始烘焙混合粉需要添加其他多种原料才能使用，不久之后，许多混合粉本身就含有鸡蛋粉等成分，免去了添加鸡蛋等辅助原料的麻烦。实际上，有些品种的混合粉还需添加好多材料，只比自己亲自动手省事一点点。

Buying Kitchenware 购买厨具

舌尖美食词汇

grater 擦菜板	**opener** 开瓶器
peeler 削皮器	**measuring cups** 量杯
spatula 铲子	**frying pan** 平底锅
cutters 模型，切割工具	**oven mitt** 烘焙手套
aluminum wrap 铝箔纸	

舌尖美食句

1.	Do you have any graters?	你们有擦菜板吗?
2.	I need a can and jar opener.	我需要一个开罐器。
3.	Can you show me how to use this peeler?	你能演示一下怎么用这个削皮器吗?
4.	These measuring cups have different size.	这些量杯有不同的大小。
5.	How much is this frying pan?	这个平底锅多少钱?
6.	Is this kind of aluminum wrap in good function?	这种铝箔纸好用吗?

7. Can you find me two spatulas?	你能给我找两个铲子吗？
8. These cookie cutters are the most popular among the housewives.	这些饼干模具最受家庭主妇们欢迎。
9. This big wok is cheaper than the small one.	这个大的炒菜锅比那个小的还便宜。
10. This pair of oven mitts suits me.	这双烘焙手套适合我。
11. What's these tools for?	这些器具是用来干什么的？
12. This coupon is not valid. It has expired.	这张优惠券无效，它已经过期了。

舌尖聊美食

Conversation 1

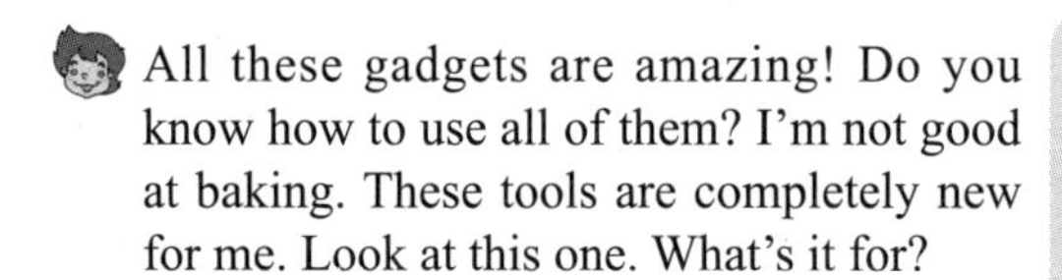

All these gadgets are amazing! Do you know how to use all of them? I'm not good at baking. These tools are completely new for me. Look at this one. What's it for?	这些小东西真神奇！你知道怎么用吗？我不太擅长烘焙糕点。这些器具，我一点儿也不熟悉。看看这个，这个是干什么用的？
 It's a sifter, used for sifting flour or sugar.	那是一个筛子，用来过滤面粉或糖。
 How about this one? Is it a smaller sifter?	这个呢？是一个小一点的筛子吗？
No. That's an egg separator. Look, I can't find the mixing bowl and the pastry brush. Could you help me with them? They must be here somewhere.	不，那是一个鸡蛋分离器。看，我找不到搅拌碗和软毛刷。你可以帮我一起找吗？它们肯定在这里某个地方。

Is this the right one?	是这个吗?
You're freaking me out. This bowl is for eating. The mix bowl I need is much bigger.	你真是让我抓狂。这是吃饭用的碗。搅拌碗要大一点。
Oh. I saw a bigger one around the corner.	噢，我在那边角落里看到一个大的。
Yes, it is. Let's go finding the pastry brush.	是，是这个。我们再找找软毛刷。
No problem. I will hunt for it from this pile of mess.	没问题。我们在这堆东西里找找。

开胃词组

be used for 被用于……

help sb. with sth. 帮助某人做某事

freak out 使发疯，使抓狂

hunt for 寻找

鲜美单词

gadget n. 小玩意，小器具

sifter n. 筛子

separator n. 分离器

bowl n. 碗

pastry n. 糕点，油酥点心

hunt v. 搜寻，寻找

pile n. 一堆

mess n. 混乱，凌乱

Conversation 2

Hi. Check out, please.	你好，请结账。
Are you a member of the Gill Supermarket?	你是吉尔超市的会员吗?
No. What would I get if I was a member? Do I have to pay to join?	不是。加入会员有什么好处？入会员要付费吗?
It's free, and you can get a coupon for every one hundred dollar purchase. What you have to do is filling out this form and bring it back here next time. The reward details are on the back of the form.	免费的，成为会员，购物价值满100美元就可获得一张优惠券。你只需填写这张表格，下次带过来。奖励具体细节在表格背面。
Thanks. I'll see about it. By the way, can I have a double bag for the kitchenware? They're quite heavy.	谢谢，我会考虑的。顺便问一下，我的厨具很重，可以用两个袋子装吗?
Sure. The charge is 95 dollars in total.	当然，一共95美元。
Can I pay by credit card?	可以刷信用卡吗?
Yes. Swipe your card here, enter your pin and sign your name, please.	可以，请刷下卡，输入密码并签名。
Here we go.	好了。
Thank you. Have a nice day.	谢谢。祝你愉快。
You too.	也祝你愉快。

开胃词组

check out 结账

have to 必须，不得不

fill out 填写

see about 考虑，安排

by the way 顺便提一下

credit card 信用卡

鲜美单词

member *n.* 成员，会员

coupon *n.* 优惠券

purchase *n.* 购买，采购

reward *n.* 奖励，奖金

detail *n.* 细节，详情

double *v.* 双倍，使加倍

kitchenware *n.* 厨房用具

charge *n.* 费用

swipe *v.* 刷（信用卡等）

How to choose cookware? 如何选购厨具?

When purchasing cookware you will need to think about the surface material, construction material, and the overall appearance and design.

选购厨具的时候，要看厨具表层的材料、构造材料、外表设计等。

Surface Material

表层材料

The most prominent cookware surface options available in the marketplace are nonstick, stainless steel, and hard anodized aluminum.

市面上，厨具的表层材料主要有不粘底的、不锈钢的和铝合金的。

Nonstick pans release food better and are easier to clean than stainless steel and hard anodized aluminum. Every home cook should have at least one good nonstick pan on their pot rack.

不粘底的厨具，相比不锈钢和铝合金，更容易让食物出锅，便于清洗。每家至少应该有一个不粘底的厨具。

Stainless steel cookware are the best choice for browning red meats, fish and poultry as well as for simmering your favorite sauces.

不锈钢厨具是煎烤红肉、鱼肉和禽肉的最佳选择，也是煨炖酱汁的好工具。

Hard anodized aluminum cookware surfaces are sealed through the anodizing process. The hardening and sealing produced by anodizing creates a cooking surface that is easy to clean and will brown meats sufficiently.

铝合金厨具表层经过阳极化处理，形成硬化的表层，方便清洗污垢，也助于有效地炒菜。

Other considerations

其他因素

Make sure that the pans are ovenproof to 400 or 500 degrees with stainless steel handles, make sure the lid is tight fitting and that the pan feels good in your hand. Look at the overall design: what design facts do you find appealing, what design aspects make the pan safer to move around, pick up and prevent burns?

挑选平底锅的时候，要选择耐温400至500度以上的，带有不锈钢手柄的，手感好的，还要确保锅盖能盖得严实。要看整体的设计：什么是自己喜欢的造型设计，而什么又是安全、实用的产品设计。

Part 3

American Cuisine
美国美食

Cuisine of the South 南部风味

舌尖美食词汇

oysters 牡蛎	**fried chicken** 炸鸡
mashed potatoes 土豆泥	**grits** 粗玉米粥
jambalaya 什锦菜	**hush puppy** 油炸玉米饼
buttermilk 酪乳	

舌尖美食句

1.	In the south, we eat oysters all the time.	在南部，我们都吃牡蛎。
2.	Here's the recipe of delicious southern fried chicken.	这是美味可口的南方炸鸡食谱。
3.	Texas specializes in barbecue and chili as well as a regional variation of Mexican food unique to Texas called Tex-Mex.	德克萨斯州以烧烤和辣椒为特色菜，还有各种墨西哥风味的食物，名叫德克萨斯-墨西哥。
4.	Orange juice is the well-known beverage of the south.	橙汁是南部著名的饮品。

5. Peaches, pecans, peanuts and Vidalia onions are very popular in Georgia.	鲜桃、胡桃、花生和维达利亚洋葱在佐治亚州很受欢迎。
6. Is grits sweet or salty?	粗玉米粥是甜的还是咸的？
7. Jambalaya is a typical food in the south.	什锦菜是南部的特色菜。
8. Buttermilk biscuits are usually served with butter, honey or fruit preserves in the south.	在南部，食用奶油饼干一般用黄油、蜂蜜或果脯作调味品。
9. It may take you two hours to make a dish of mashed potatoes.	做一道土豆泥也许要花你两个钟头的时间。
10. Some southern meals consist of only vegetables with no meat dish at all.	南部的一些菜肴只有蔬菜而没有肉。
11. I prefer the sweet hushpuppy.	我喜欢吃甜甜的油炸玉米饼。
12. Would you like to try some beans and greens?	你想尝一些豆类和蔬菜吗？

舌尖聊美食

Conversation 1

Cathy, what's the course of cornmeal called?	凯茜，这玉米粉做的饼叫什么呀？
Oh, it's hush puppy.	叫油炸玉米饼呀。
Hush puppy? Sounds interesting. Why does it contain "puppy" in its name?	油炸玉米饼？挺有意思的。那为什么它的名字里有"小狗"这个词呢？

I heard one version of the story. There's an African cook in Atlanta. One day, when he was frying cornmeal and catfish, a puppy began to howl. In order to make the puppy quiet, he gave it a bowl of cornmeal and ordered, " Hush, puppy." Then, the name becomes popular.

我听说过这样一个故事：一天，亚特兰大的一位非洲厨师在油炸玉米饼和鲶鱼的时候，小狗嚷叫起来，他给小狗一碗玉米饼，然后下令"安静，小狗"。从此这个名字就流传开来了。

Wow, how interesting it is.

哇，真有意思。

Oh, yeah. But I think "hush puppy" is better to be named "hush billy" .

嗯，但是我觉得这道菜叫"安静，比利"更合适。

Why?

为什么?

That's because you always feel excited and lick your lips when I fry hushpuppies.

因为每次我油炸玉米饼的时候，你总是非常兴奋，垂涎三尺。

Hah, that's true.

哈哈，没错。

开胃词组

in order to 为了

be named 被命名为

lick one's lips 舔嘴唇，流口水

鲜美单词

course *n.* 一道菜

cornmeal *n.* 玉米粉，玉米片

contain *v.* 包含，容纳

version *n.* 版本；说法，描述

catfish *n.* 鲶鱼

order *v.* 命令，下令

lick v. 舔

lip n. 嘴唇

Conversation 2

Hi, Sam. Do you come from the south of the USA?	嗨，山姆，你是来自美国南部的吗？
Yes, I'm from New Mexico where is known by its gourmet food.	是的，我来自新墨西哥州，那里因遍地美食而闻名。
Really? Could you tell me more about it?	真的吗？能给我介绍一下吗？
Sure. First of all, I must talk about the grits, which are usually served for breakfast. Grits refer to a ground-corn. Modern grits are commonly made of alkali-treated corn known as hominy. Grits are prepared by simply boiling the ground kernels into porridge until enough water is absorbed or vaporized to leave it semi-solid.	当然。首先，我得给你介绍粗玉米粥，一种常见的早餐食物。粗玉米就是指玉米粉。现在的玉米粥就是由玉米做的玉米片粥，放水煮，玉米粒变成粥状，水被吸干，最后变成糊状。
Wow, it sounds delicious.	哇，听起来很好吃.
That's right. It's also good for your health. And secondly, fried chicken is the best known cuisine among this region. The chicken is deep fried in fat.	没错，这种食物还有利于身体健康。其次，我要说的就是，炸鸡块是南部著名的美食。鸡肉放在油锅里炸熟。
It must smells good and tastes yummy.	那闻起来肯定很香、很好吃。
The southern people also like to eat seafood, such as seafood boil. It involves shrimp, and many kinds of shellfish, like clam, and so on.	南部的人们还喜欢吃海鲜，例如海鲜火锅。里面有虾，还有许多贝壳，像蛤蜊什么的。

Thank you. I think I would like a trip to the south of the United States.

谢谢你介绍了那么多。我真想来一次美国南部之旅。

开胃词组

come from 来自于
first of all 首先
refer to 涉及，指的是
made of 由……做成的
and so on 等等；诸如此类

鲜美单词

gourmet *adj.* 美食的，美味佳肴的
grit *n.* 粗玉米粉
serve *v.* 给……提供，上菜
hominy *n.* 玉米粥
kernel *n.* 核，粒
porridge *n.* 粥，麦片粥
absorb *v.* 吸收
vaporize *v.* 使蒸发
cuisine *n.* 菜，菜肴
region *n.* 地区，地域
clam *n.* 蛤，蚌，蛤蜊

The states in the south are Alabama, Arkansas, Florida, Georgia, Kentucky, Louisiana, Mississippi, North Carolina, Oklahoma, South Carolina, Tennessee, Texas, Virginia, and West Virginia. Some people also consider Maryland part of the "South."

美国南部的州包括亚拉巴马州、阿肯色州、佛罗里达州、佐治亚州、肯塔基州、路易斯安那州、密西西比州、北卡罗来纳州、俄克拉荷马州、南卡罗来纳州、田纳西州、德克萨斯州、弗吉尼亚州、西弗吉尼亚州。有的人也称马里兰州为南部的一部分。

The population of the Southern United States is made up of many different peoples who came to the region in a variety of ways, each contributing to what is now called "Southern cooking." American Indians, native to the region, taught European settlers to grow and cook corn, a grain unknown in Europe at the time. Spanish explorers in the 1500s brought pigs with them, introducing pork to the region West Africans carried some of their traditional foods with them, such as watermelon, eggplant, collard greens, and okra, when they were brought to the United States by force as slaves beginning in the 1600s. Creoles, known for their unique use of spices, are descended from French and Haitian immigrants who later mingled with Spanish settlers in the New Orleans area. "Cajuns," also recognized for their unique style of cooking, were originally Acadians, French settlers in Nova Scotia who were driven out by the British in 1755 and made their way to New Orleans. In Louisiana, crawfish (resemble miniature lobsters) and catfish are popular, prepared in dozens of different ways. Fried catfish is popular all across the South. Texas's spicy and flavorful "Tex-Mex" cuisine reflects the state's close proximity to the spicy cuisine of Mexico.

美国南部人口由来自世界不同地区的人组成，因此形成独具特色的南部美食风味。印第安人土著居民给欧洲移民传授玉米的种植方法，在当时的欧洲，人们都没见过这种植物。16世纪西班牙探险者们带来了圈养猪，猪肉成为家常美食。17世纪，非洲人被送往美洲做奴隶时，也带来他们的传统食物，比如西瓜、茄子、甘蓝叶，还有秋葵荚。克里奥耳人喜欢吃辣味，由法国人和海地人带入美国，世代相传，并与新奥尔良地区的西班牙移民者的食物融合。卡津风味由阿卡迪亚人创造，他们是法国移民者的后裔，1755年被英国人驱赶到新奥尔良地区，他们的食物也因此自成一派。在路易斯安那州，小龙虾和鲶鱼非常受欢迎，有不同的烹饪方式。油炸鲶鱼在南部大受欢迎。德克萨斯州的“德克萨斯-墨西哥”辛辣风味，与墨西哥辣味有千丝万缕的联系。

读书笔记

Cuisine of the North-Eastern 东北部菜肴

舌尖美食词汇

bagel 硬面包圈	**pastrami** 熏牛肉
submarine sandwiches 潜艇三明治（矩形三明治）	**Buffalo wings** 布法罗鸡翅
sponge toffee 海绵奶糖	**maple float** 枫糖冰浮
pretzel 椒盐脆饼	**clam chowder** 蛤肉杂烩

舌尖美食句

1. In the 17th century, Native Americans and English immigrants came into contact along New England's rocky coast and helped define Northeastern regional cooking.

17世纪，美国土著人和英国移民者在新英格兰海岸相处，形成了东北部地区烹饪特色。

2. Pizza, bagels, pastrami and submarine sandwiches are popular in New Jersey.

比萨、硬面包圈、熏牛肉，还有潜艇三明治在新泽西很受欢迎。

3. New York City is especially famous for its Italian and Chinese cuisines.

纽约有著名的意大利美食和中国风味美食。

4. Buffalo is known for its Buffalo wings, sponge toffee and pastry heart.	布法罗有著名的布法罗鸡翅、海绵奶糖和糕点。
5. Popover is a type of small bread that is empty inside.	蓬松饼是一种内部中空的小面包。
6. Binghamton is the home of the spiedie, a unique type of sandwich.	宾厄姆顿是司皮迪的发源地，司皮迪是一种当地特有的三明治。
7. Milk, pure maple syrup and vanilla ice cream are the basic ingredients for Vermont maple float.	牛奶、纯枫糖及香草冰淇淋是佛蒙特枫糖冰浮的基本原材料。
8. Pretzels are a common snack in Pennsylvania.	椒盐脆饼是宾夕法尼亚州常见的小吃。
9. How about order Rhode Island's clam chowder, which unlike New England clam chowder?	点一些罗得岛州的蛤肉杂烩怎么样？它与新英格兰的蛤肉杂烩风味不同。
10. Vermont is famous for its maple syrup and related products such as maple candy.	佛蒙特州因其枫糖浆及枫糖等相关产品而闻名。
11. The Northeast played a major role in establishing America's restaurant culture.	东北部在美国餐馆文化中扮演重要角色。

舌尖聊美食

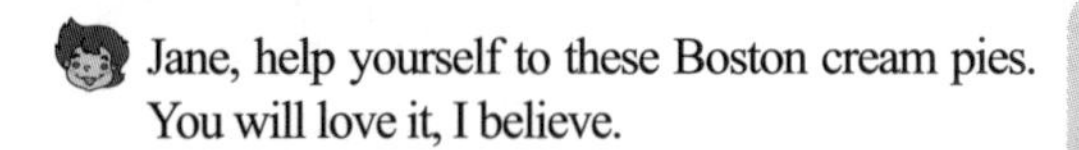

Conversation 1

Jane, help yourself to these Boston cream pies. You will love it, I believe.	简，你尝尝这些波士顿奶油派。相信你会喜欢。
Thank you. It tastes very nice. But I'm confused. What did you call it just now?	谢谢。真好吃。但是我有点困惑。你刚才说的，这个叫什么名来着？

Boston cream pies.	波士顿奶油派啊。
But I think they're cakes, not pies. Pies are wedge-like.	但是它们看起来就是蛋糕啊，不是派。派是那种楔形的。
Oh, yeah. They look like cakes, but we call them pies, because the dividing line between a cake and a pie is vague. Probably, the first versions might be baked in pie tins, as pie tins were more common than cake pans.	嗯，没错。它们看起来是像蛋糕，但是我们却称之为派，因为蛋糕和派的区别很模糊，也许，第一个是在派罐里做的吧，因为派罐比蛋糕盘更常见。
Oh, I got it. Pies hand down their names from their originality.	明白了。"派"的称呼是从它的诞生方式中产生的呀！
Yes. I'd like to tell you more stories behind the traditional food.	是的，我还会告诉你更多美食背后的小故事呢。
Great! Thanks.	太棒了。谢谢。

开胃词组

help yourself to 请自便，随便吃

just now 刚才，刚刚

dividing line 分界线，界限

hand down 传递，传给（后代）

鲜美单词

cream *n.* 奶油，乳脂

confuse *v.* 使困惑，使弄糊涂

wedge *n.* 楔；楔形物

vague *adj.* 模糊的，不明确的

tin *n.* 烤模，烤盘

common adj. 常见的，普通的

originality adj. 首创的，独创的

traditional adj. 传统的

Conversation 2

Mom, how do you make the tasty maple float?	妈妈，那么好吃的枫糖冰浮，你是怎么做的?
Well, it's very easy. What you need to do is pour and mix.	不难做啊。你要做的只是把材料倒进去，搅拌搅拌就行了。
Really?	真的吗?
Look, you get the ingredients ready and prepare a mix bowl.	看，你先准备好原料，还有一个搅拌碗。
I see. Milk, pure maple syrup and vanilla ice cream.	明白。准备好原料：牛奶、纯枫糖，还有香草冰淇淋。
Yes. First, pour 6 ounces of milk into the mix bowl and stir in maple syrup. One tablespoon of syrup would be fine. Then, mix them well, and top them with a scoop of ice cream. That's it!	对，首先，将6盎司的牛奶倒入搅拌碗里，并加入枫糖搅拌。一汤匙的枫糖就可以了。然后，搅匀。再点缀一勺冰淇淋。完成了！
Hah, it's amazing and magical. I'll try it.	哈，真神奇！我也来试试。
Yeah, go ahead.	嗯，那你试试。

开胃词组

maple float 枫糖冰浮

get ready 准备

pour into 倒入，倒进

top with 用……覆盖在上面

go ahead 着手做，开始进行

鲜美单词

pour v. 倾倒，倒出

mix v. 混合

ingredient n. 原料

prepare v. 准备，预备

ounce n. 盎司

stir v. 搅拌

tablespoon n. 大汤匙

scoop n. 勺，勺状物

magical adj. 奇妙的，神奇的

There is some dispute about who came up with the original hot wing appetizer - Buffalo Wings, but most credit the Anchor Bar in where else but Buffalo, New York, USA.

关于布法罗鸡翅的起源，有一些争议，但大多数人认为布法罗鸡翅起源于美国纽约州布法罗的船锚酒吧。

The historic creation date for Buffalo Wings was October 30, 1964, when owner Teressa Bellissimo was faced with feeding her son and his friends a late snack. Having an excess of chicken wings on hand, she fried up the wings, dipped them in a buttered spicy chile sauce, and served them with celery and blue cheese dressing as a dipping sauce to cut the heat. The wings were an instant hit.

布法罗起源于1964年10月30日，船锚酒吧老板泰瑞莎·贝里西莫给她的儿子及他的朋友们做夜宵。由于鸡翅过剩，她油炸这些鸡翅，用黄油辣酱做蘸酱，再配以芹菜和蓝纹奶酪做调味品。这种鸡翅顿时风靡起来。

The city of Buffalo has designated July 29 as "Chicken Wing Day," and today, the Anchor Bar serves up more than 70 thousand pounds of chicken per month! The Anchor Bar original recipe for hot sauce is now sold commercially.

布法罗市把7月29日定为"鸡翅庆祝日"，如今的船锚酒吧每月出售7万磅的鸡翅。该酒吧的原始酱汁配方也出售。

Many restaurants across the United States soon jumped on the chicken wing bandwagon. In fact, it is difficult to find a restaurant that does not carry some version of chicken wings on the menu. Many have also come up with different flavors for chicken wings, ranging from jerked wings to Oriental flavors. So even if you can't handle the hot stuff, there are chicken wing recipes for you.

美国许多饭馆开始推崇鸡翅这道菜。实际上，大多数很难发它们各有什么不同。但是也有不同口味的，有重口味的，东方清淡口味的，所以如果吃不了辣的，还有其他的口味可供选择。

Cuisine of the Western 西部美食

UNIT 03

舌尖美食词汇

gooey buttercake 牛油蛋糕	**cranberry salsa** 蔓越莓辣酱
pemmican 干肉饼	**broiled salmon steak** 汁烤三文鱼扒
apple crisp 苹果酥	**blueberry muffin** 蓝莓松饼
salmon 三文鱼	**marinara sauce** 意式番茄酱

舌尖美食句

1.	How does the gooey butter cake taste?	牛油蛋糕吃起来味道怎样？
2.	I have ordered some broiled salmon steak.	我点了汁烤三文鱼扒。
3.	Cranberry salsa is always requested at gatherings.	蔓越莓辣酱常在聚会中食用。
4.	How to make such delicious pemmican?	这么美味的干肉饼是怎么做的？
5.	She's crazy about apple crisp.	她特别喜欢吃苹果酥。

6. Hot smoked salmon is his special dish.

热熏三文鱼是他的拿手好菜。

7. Chinese immigrants first introduced the fortune cookie to California.

华人首次把幸运饼干带到了加利福尼亚。

8. What are the typical dishes in the West?

西部的特色菜是什么?

9. How do you make a blueberry muffin?

你怎么做蓝莓松饼?

10. In the Northwest, salmon is involved in various specialties.

在西北部，三文鱼成为各色菜的食材。

舌尖聊美食

Conversation 1

Excuse me. I have no idea about the strange name of these dishes. Could you recommend some local appetizers for me?

打扰一下。我不了解这些千奇百怪的菜名。你能推荐一些本地特色菜吗?

OK. St. Louis ravioli is our best seller. It's very popular in St. Louis. Here, on the first page of the menu.

好的。圣路易斯方形饺是我们的热门菜。在圣路易斯地区很受欢迎。这里，菜单的第一页就是。

It's deep fried. What's in it?

它是油炸的。里面都有什么?

Generally, some meat or cheese could be wrapped in it. Then breaded and deep fried. That will be perfect, if you dip it in marinara sauce.

通常里面包上一些肉或奶酪，然后裹上面包屑，再油炸。如果蘸着意大利番茄酱食用，会更完美。

So, it is crispy, isn't it?

看来它是脆的，是吧?

 Yes. Would you like some? | 是的。你想来点吗？

 Wonderful. I'd like some. | 太好了。给我来点。

开胃词组

have no idea 不知道，不了解

best seller 热销，畅销

be popular 受欢迎

鲜美单词

strange adj. 陌生的，生疏的

dish n. 菜，菜肴

recommend v. 推荐，建议

appetizer n. 开胃菜，开胃品

ravioli n. 意大利方饺

menu n. 菜单

marinara n. （意大利的）海员沙司

crispy adj. 脆的，酥脆的

Conversation 2

 Hi, Linda. My son is crazy about sourdough pancakes, and he asks me to make pancakes every single day. | 嗨，琳达。我的儿子特喜欢吃美式煎饼，他每天都叫我做煎饼。

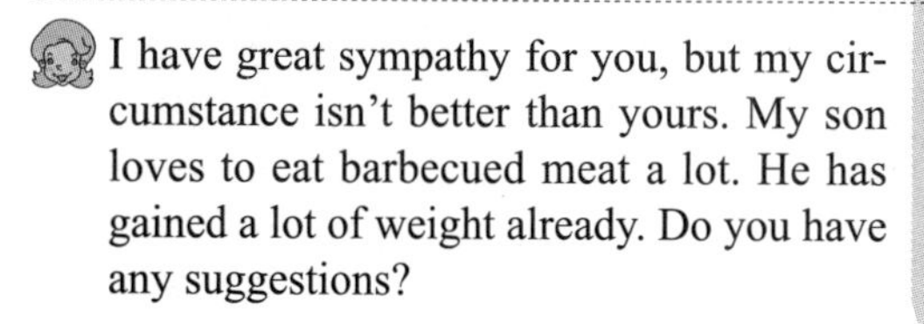

I have great sympathy for you, but my circumstance isn't better than yours. My son loves to eat barbecued meat a lot. He has gained a lot of weight already. Do you have any suggestions? | 虽然我很同情你，但是我的状况也好不到哪里去。我的儿子喜欢吃烤肉。他现在已经长胖了许多。你有什么建议吗？

I think the kids should eat more healthy food. As I know, a lot of cuisines here are oily or fatty, such as roast rib, grilled lamb and fried pancake.	我觉得孩子们应该多吃健康的食物。我知道，这里的美食都是油腻的，比如烤牛排、烤羊肉和煎饼。
Yes, that's right. I think the seafood is good for them.	是的，没错。我觉得海鲜对他们来说不错。
Oh, good idea. Broiled salmon steak, hot smoked salmon, pan-fried smelt are very popular here and low in fat as well.	哦，好主意。汁烤三文鱼扒、热熏三文鱼和生煎多春鱼在这里很受欢迎，脂肪含量也很低。
Well, these specialties are delicious. The kids must love them and get away from the pancake or barbecued meat.	嗯，这些特色菜也十分美味，孩子们肯定喜欢吃，他们就远离那些煎饼或烤肉什么的了。

开胃词组

be crazy about 对……痴迷，迷恋

every single day 每一天

gain weight 增重，增肥

such as 例如

get away from 远离，摆脱

鲜美单词

sourdough *n.* 酵母

sympathy *n.* 同情，怜悯

circumstance *n.* 环境，境遇

weight *n.* 重量，体重

cuisine *n.* 菜肴，饭菜

roast *v.* 烤，烘

grill v. 烧烤

broil v. 烤，焙

specialty n. 特色菜，招牌菜

The cuisine of Western United States is distinct from that of the rest of the country. In the Northwest, salmon is involved in various specialties. In the Plain States, such as Utah, beef is a major component of the ingredients and other wild food is part of the cuisine as well. In the Rocky Mountain, oysters cater for the need of some tourists. Along the West Coast, wine and seafood become an important industry for national development. On the border area, the culture of Mexican and other South American countries influence the cuisine of the West, and Latino food is increasingly seen through this part of the US. In areas where sheep ranching is important, family-style Basque cuisine is popularized. In a word, dishes vary as you roam around the Western land.

美国西部的菜肴和美国其他地区的风格有所差异。在西北部，三文鱼成为各式餐点的主要食材。在诸如犹他州等平原州，牛肉是主要材料，其他野生食物同样构成了当地美食的一部分。在落基山脉，牡蛎是许多游客的最爱。在西部海岸区，葡萄酒和海鲜已经成为国民经济发展的重要组成部分。在边界地区，墨西哥文化和其他南美国家文化影响了美国西部的饮食习俗，拉丁美洲食物在美国西部日趋常见。在绵羊养殖为主的地区，以家庭畜牧业为特色的巴斯克饮食文化盛行。总之，当你漫步在美国西部地区时，餐饮习俗也随着你前行的脚步悄然发生变化。

Great Lakes Regional Cuisine 湖区菜

舌尖美食词汇

cheese 奶酪	**Chicago mix popcorn** 芝加哥爆米花
kringle 双环面包圈	**sugar cream pie** 奶油糖派
limburger 林堡干酪	**Turducken** 特大啃（由火鸡、鸭肉或鸡肉制成）
hot-smoked salmon 热熏三文鱼	**Cobb salad** 科布沙拉（一种熏肉和牛油果的沙拉）

舌尖美食句

1. Milk, butter and cheese are staples in the Great Lakes diet.	牛奶、黄油、奶酪是湖区的主要食品之一。
2. Chicago mix popcorn is preferred in social events.	芝加哥爆米花是社交活动中受人喜爱的食品。
3. Kringle is a kind of Danish dessert.	双环面包圈是一种丹麦风味的点心。
4. You can serve sugar cream pie warm, but I like it better cold.	你可以热着吃奶油糖派，但我喜欢吃凉的。

5. Wisconsin is supposed to have a good limburger sandwich.	威斯康星州应该有好吃的林堡干酪三明治。
6. Turduchen is a dish consisting of a deboned chicken stuffed into a deboned duck, which is in turn stuffed into a debond turkey .	特大啃是一道无骨鸭肉里塞无骨鸡肉再将其塞入无骨火鸡的菜。
7. I got hooked on hot-smoked salmon.	我特别爱吃热熏三文鱼。
8. You can buy a bowl of St. Louis raviolis at a discount today.	你今天购买圣路易斯方饺可以打折。
9. Cobb salad is my favorite side dish.	科布沙拉是我的最喜欢的配菜。
10. Pork is a common ingredient in Great Lakes cooking.	猪肉是湖区烹饪常用的食材。

舌尖聊美食

Conversation 1

How's this chicken booyah?	这道比利时鸡汤尝起来怎么样?
It's special and delicious. Do you know about the history of chicken booyah?	味道很特别，很美味。你知道这个鸡汤的历史吗?
Of course. Booyah is lovingly called "Belgian Penicillin" . The first Belgian immigrants arrived in Wisconsin in 1853. They had their own language called "Walloon". They introduced chicken booyah into United States. It is believed that the word "Booyah" comes from the word "bouillon".	当然。布哈有个可爱的称呼"比利时青霉素"。1853年比利时人移民美国威斯康星州。这些比利时人有自己的语言"瓦隆语"。他们把布哈鸡汤介绍给人们。据说"布哈"这个词来源于"肉汤"这个词语。

 It's interesting. Is it easy to make this soup?

真有意思。做这道菜很简单吧？

 Well, that's not easy. Instead, it's a large project. This chicken soup is typically made in large 10 or 20-gallon batches, cooked outdoors over a wood fire and worked on by several people at once.

嗯，那可不简单，反而是一个大工程。这种鸡汤得用容量达到10到20加仑的炉子在户外火堆上煮，需要好几个人一起做。

 Oh, it sounds quite complicated.

噢，听起来很复杂。

 And you should prepare a lot of ingredients, such as roasting chicken, beef, pork, onions, carrots, potatoes, fresh peas, ground pepper and small bunch of celery.

还有，你需要准备许多食材，比如烤鸡、牛肉、猪肉、洋葱、胡萝卜、土豆、鲜豆、辣椒粉和一些芹菜。

 It's not a piece of cake, is it?

这可不是小事一桩，对吧？

 No, it's obviously not.

对，那可不是嘛。

开胃词组

arrive in 抵达，到达

of course 当然，肯定

work on 从事……；继续工作；致力于

at once 一起，同时

a piece of cake 小事一桩

鲜美单词

delicious adj. 美味的。可口的

Belgian n. & adj. 比利时人；比利时人的

immigrant n. 移民，侨民

instead adv. 反而，相反地

project *n.* 工程，项目

batch *n.* 一炉，一群

outdoors *adv.* 在户外

bunch *n.* 一串，一堆

celery *n.* 芹菜

obviously *adv.* 明显地

Conversation 2

Excuse me. I have ordered two kringles, but you served me with two paczkis.	对不起，我点了两个双环面包圈，你给我的却是两个波兰甜甜圈。
Oh, I'm sorry. Let me have a check. What's your table number?	哦，对不起。我核实一下。你们的桌号是多少？
We take number 122. Here's the label.	我们坐的是122号。看，这是桌牌。
Oh, I have to apologize that I wrote down the wrong table number. I will change the dish for you as soon as possible.	噢，我向你们道歉，我记错号码了。我马上给你们换菜。
Will it be done in 10 minutes? We are leaving in no more than 15 minutes. Please cancel this order if the food would not be ready in 10 minutes.	10分钟之内能做完吗？我们15分钟之后就要走了。如果10分钟以内没做好，就不要上了。
OK. I'll be right back.	好的。我马上回来。

开胃词组

write down 写下，记下

as soon as possible 尽快地，尽可能地

no more than 至多，不超过

鲜美单词

order v. 点菜

check v. 检查，核对

serve v. 招待，服务

label n. 标签，标记，符号

apologize v. 道歉

change v. 改变；更换，替换

cancel v. 取消，撤销

Foods of the Great Lakes Region 美国湖区美食

The Great Lakes region was originally populated by American Indians who taught later European settlers how to hunt the local game, fish, and gather wild rice and maple syrup, as well as how to grow and eat corn and native squashes and beans.

美国湖区人口原来主要以北美印第安人为主，他们向欧洲移民者传授打猎、捕鱼、收集野生水稻和枫糖浆等技能，还教授他们如何种植及食用玉米，南瓜和黄豆。

The European immigrants, mostly from Germany, Scandinavia, Holland, Poland, and Cornwall, England, each shared their traditional dishes with the rest of America. The Germans contributed frankfurters (hot dogs), hamburgers, sauerkraut, potato salad, noodles, bratwurst, liverwurst and pretzels to the American diet. Scandinavian foods include lefse (potato flatbread), limpa, lutefisk, and Swedish meatballs, as well as the smorgasbord. The Polish introduced kielbasa (a type of sausage), pierogies (a type of stuffed pasta), Polish dill pickles and babka. Pancakes are a

Dutch contribution, along with waffles, doughnuts, cookies and coleslaw. Miners from Cornwall brought their Cornish pasties and small meat pies that were easily carried for lunch.

欧洲移民者大部分来自德国、斯堪的纳维亚半岛、荷兰、波兰，还有英格兰的康沃尔，他们与本土居民分享他们的传统美食。德国人带来了熏肠（热狗）、汉堡、腌菜、土豆沙拉、面条、小香肠、猪肝香肠、椒盐脆饼。斯堪的纳维亚半岛的美食包括挪威煎饼（土豆面饼）、甜黑麦面包、咸鱼，还有瑞典的肉丸、大杂烩。波兰人引入了熏肠、馅饼（一种带馅的面食）、波兰泡菜、巴布卡蛋糕。荷兰人献出他们的薄煎饼、华夫饼、甜面圈、曲奇饼，还有生菜沙拉。康沃尔的矿工带来他们的肉馅饼、午餐肉饼。

Later immigrants from Arab countries settled in Detroit, Michigan, and introduced America to foods like hummus (pureed chickpeas), felafel (deep-fried bean cakes) and tabbouleh (bulgur wheat salad).

阿拉伯人来到密歇根州的底特律市安定下来，与当地人分享他们的特色食物，如鹰嘴豆泥、豆泥球，还有塔博勒沙拉。

读书笔记

California Specialties 加州特色菜

舌尖美食词汇

cheeseburger 奶酪汉堡	**tamale** 玉米粉蒸肉
cheesy asparagus pizza 奶酪芦笋比萨	**carne asada** 墨西哥扒牛柳
cioppino 海鲜番茄杂烩	**sope** 饼包肉
pozole 墨西哥辣酱浓汤	**shrimp tostada** 虾仁玉米饼

舌尖美食句

1. Mexican and Spanish-origin cuisine is very influential and popular in California, particularly Southern California.	墨西哥风味和西班牙风味对加州饮食影响非常深远，尤其加州南部地区。
2. Seafood is an important staple in the average California diet.	海鲜是加州常见的重要主食之一。
3. The "Double-Double" cheeseburger is impressing in California.	加州的双芝士汉堡真是令人印象深刻。
4. What's the price of tamales?	玉米粉蒸肉多少钱?
5. The cheesy asparagus pizzas are really tasty.	奶酪芦笋比萨真的很美味。

6. I don't think I have any more space for carne asada. May I have a cup of milk?	我恐怕吃不下更多的墨西哥扒牛柳了。我可以来杯牛奶吗?
7. Cioppino is related to various regional fish soups and stews of Italian cuisine.	海鲜番茄杂烩里有特色鱼汤和意式炖肉。
8. Do you want to try some sope?	你要尝一下饼包肉吗?
9. Pozole is a Mexican broth-based soup made with meat and hominy, a type of corn.	墨西哥辣酱浓汤是一种用肉和玉米制作的墨西哥肉汤。
10. This restaurant serves authentic Mexican food.	这家餐馆有正宗的墨西哥美食。
11. California rolls are packed with savory crab and creamy avocado.	加州卷里包着香薄荷虾肉和奶油牛油果。

舌尖聊美食

Conversation 1

 Hi, Sean. Do you like California roll?	嗨，肖恩。你喜欢加州卷吗?
 Well, I have seen it in the pictures, but I have never eaten it before.	嗯，我看过它们的图片，但从来没尝过。
 Really? I have packed some California rolls for my lunch. I would like to share them with you. Look, here they are.	真的吗? 我带了些加州卷作为午餐。我想跟你一起分享。看，这就是。
 Wonderful. I guess they look a little different from Japanese sushi.	棒极了。我认为它们看起来与日本寿司不一样。

You're right. Americans like to wrap the nori inside, but the Japanese usually make the nori outside.	没错。美国人把海苔包在加州卷的里面，而日本的寿司的海苔是在外面的。
Why do they make a change for that?	为什么要做这样的改变?
We aren't used to the nori outside because it's not quite easy to chew. And Americans use crabmeat or shrimp meat instead of sashimi.	我们美国人不习惯把海苔包在外面，因为不好嚼。还有，美国的加州卷用蟹肉或虾仁做肉馅而不用生鱼片。
So it's tailored according to American's desire.	所以，加州卷根据美国人的口味，做了改良。
Exactly.	没错。

开胃词组

share with sb. 与某人一起分享

different from 不同于，异于

make a change 更改，改变

be used to 习惯于

according to 根据，按照

鲜美单词

pack *v.* 打包，包裹；携带

roll *n.* 小圆面包，面包卷

sushi *n.* 寿司

wrap *v.* 缠绕

nori *n.* 海苔

chew *v.* 咀嚼

tailor *v.* 裁剪；调整使适应

desire *n.* 欲望，渴望

Conversation 2

What a fancy place for eating! They said this restaurant serves the most authentic California flavor in the US.

这里真是吃饭的好地方。他们说这里有全美最正宗的加州风味。

Yes. Cobb salads, seared scallops and carne asada fries are the best sellers in this restaurant.

是的。科布沙拉、香煎扇贝还有卡尔墨西哥扒牛柳薯条是这家餐馆的热销菜。

All right. We make the seared scallop first. I think you prefer Mexican flavor. Have you been accustomed to the California cuisine?

好的。我们先点香煎扇贝吧。我认为你更喜欢墨西哥风味的吧。你现在习惯加州美食了吗?

No problem. Many years ago, the immigrants from Mexico brought their traditional food to California. Look, burritos, refried beans and tortas are all Mexican traditional food, and they are very common here.

没问题。许多年前,墨西哥移民把他们的传统美食也带到了加州。看,墨西哥卷、墨西哥豆泥,墨西哥三明治都是传统的墨西哥风味,在这里很常见的。

Sounds great. Why don't we order some Mexican food? Now, it's your turn.

听起来不错。我们为何不点一些墨西哥菜呢?现在,轮到你点菜啦。

OK. Let me take a look.

好的,我看看。

开胃词组

best seller 畅销

be accustomed to 习惯于

traditional food 传统食品

it's one's turn 轮到某人

鲜美单词

fancy *adj.* 昂贵的，奇特的

authentic *adj.* 真正的，准确的，可靠的

Cobb *n.* 科布，科布人

sear *v.* 烧焦

scallop *n.* 扇贝，贝壳

Mexican *adj.* 墨西哥的，墨西哥人的

accustom *v.* 使习惯

immigrant *n.* 移民，侨民

burrito *n.* 墨西哥玉米煎饼

torta *n.* 大圆形糕饼

traditional *adj.* 传统的

舌尖美食文化

The cuisine of California 加州特色菜

The modern cuisine of California is majorly influenced by French Nouvelle cuisine, Latin American cooking, Mediterranean culinary styles and Asian cuisine.

加州的现代烹饪主要受到法国新式烹饪、拉丁美洲烹饪、地中海烹饪风格和亚洲烹饪的影响。

Car restaurants, often called the drive-thru restaurants, form the legendary Southern California food culture, which are automobiles selling fast foods, burgers and sandwiches. Gourmet burgers are very popular in the state.

汽车快餐也称为免下车餐馆，形成独特的南部加州饮食文化，主要向车主们出售套餐、汉堡和三明治。美味的汉堡在加州很受欢迎。

Barbecued foods on the other hand have a special place in Californian cuisine, as the Mexicans used to cook beef in ranched pit barbecues during the 1840s. However, barbecued dishes in this state also imbibed the Southwestern American cooking styles of Oklahoma, Arizona, Texas and New Mexico. Pork baby back ribs, beef ribs, steaks and sausages are on the top in this category.

另一方面，烧烤在加州美食中也占有一席之地，据说墨西哥人早在19世纪40年代就喜欢在农场里烤牛肉。然而，加州的烧烤也融合美国南部一些地方的特色，如俄克拉荷马州、亚利桑那州、德克萨斯州和新墨西哥州。烧烤乳猪排、小牛排、牛排和香肠最受欢迎。

读书笔记

UNIT 06 Delicious Fast Food 美味快餐

舌尖美食词汇

take-out 外卖	**drive-in restaurant** 汽车餐馆
Big Mac hamburger 麦当劳的巨无霸汉堡	**Dunkin's Donuts** 邓肯甜甜圈
chips 薯条	**Taco Bell** 塔可钟
sandwich 三明治	**hotdog** 热狗

舌尖美食句

1. The United States has the largest fast food industry in the world.
美国拥有世界上最发达的快餐行业。

2. Fast food outlets are take-away or take-out providers, often with a "drive-through" service.
快餐店提供外卖，通常提供“免下车”快餐服务。

3. Drive-in restaurants were introduced in 1920s.
汽车餐馆诞生于20世纪20年代。

4. McDonald’s unsurprisingly takes the prize for the biggest fast food chain across the U.S.
毫无疑问，麦当劳赢得“美国最大的连锁快餐”称号。

5.	Sales of Dunkin's Donuts increased steadily last year.	邓肯甜甜圈去年销售量稳定上涨。
6.	When do you usually eat hotdog?	你通常什么时候吃热狗？
7.	A cheeseburger is a hamburger topped with cheese.	芝士汉堡就是一种在上面涂有奶酪的汉堡。
8.	Would you like some soda drink?	你来点苏打饮料吗？
9.	Do you know how to make a sandwich ?	你知道怎么做三明治吗？
10.	I have ordered pizza from Papa John's.	我已经在棒约翰点了比萨。
11.	A bunch of crazy new products have hit the menus in Taco Bell, including the famous Doritos Locos Tacos.	塔可钟有许多奇怪的新品上市，包括著名的多力多滋薯片玉米卷。
12.	The White Castle hamburger chain was probably the first burger bar.	白色城堡汉堡连锁可能就是首家汉堡快餐店。

舌尖聊美食

Conversation 1

People say that Americans like nothing apart from "fast food", especially the young people. Do you think this is true? And what do you think of fast food?	人们都说美国人离不开快餐，尤其那些年轻人，你觉得对吗？你对快餐有什么看法？
I don't have anything against it, really! It's OK! I mean, you hear people saying fast food are all junk food, but most of the time they taste good. Though I wouldn't want to live off fast food all the time!	我不反对这个说法。没问题。我意思是说，人们常说那是垃圾食品，但快餐吃起来真的很美味。我觉得我离不开快餐。

I think fast food tends to make you fat. The stuff from McDonald's or KFC contains too much fat and calories.	我觉得快餐容易让人发胖。麦当劳和肯德基的食物含有太多脂肪和热量。
Hamburgers are all right. I mean, my mum says it's all junk, but frankly I can't really see what makes it any different from the stuff she cooks. I mean hamburgers fill you up; and that's what the food is supposed to do, isn't it? There's meat, vegetables and bread and cheese; as far as I'm concerned, that's a pretty balanced diet.	汉堡还可以。我是说，我妈妈说汉堡是垃圾食品，但是坦白讲，我觉得跟她在家做的饭其实没什么两样。我的意思是汉堡一样让人填饱肚子，这就是食物的用处，对吧？它里面有肉、蔬菜、面包还有奶酪，我觉得那还真是一个营养丰富的餐点呢！
Anyway. What's the most popular fast food in the United States?	不管怎样，在美国什么快餐最受欢迎啊？
All sorts of stuff. Burgers, fries and soft drinks are typical fast food in America. Cheeseburger, hotdog, French fries are really popular. And I think pizzas, fish fingers, lasagna, things like that, also belong to fast food.	太多了，各种各样的东西。汉堡、炸薯条，还有软饮料是美国的典型快餐搭配。芝士汉堡、热狗、薯条都非常受欢迎。还有比萨、炸鱼条、烤宽面条等，我觉得也都属于快餐。
OK. The fast food restaurants can be seen everywhere in America. Is that true?	知道了。快餐店在美国遍地都是，对吗？
Yeah. The fast food chains are well developed in United States. McDonald's, Subway, Burger King, Pizza Hut, Starbucks are top on the list.	对。在美国，快餐连锁店非常发达。麦当劳、赛百味、汉堡王、必胜客和星巴克都是排名前列的快餐店。
They must be in fierce competition.	他们竞争肯定也很激烈。
Absolutely. Many advertisement signs stand by the roadside.	当然了。路边布满了这些快餐广告牌。

开胃词组

apart from 脱离，除此……之外

fast food 快餐

tend to 趋向，往往易于

fill up 充满，填满

as far as 据……，就……

鲜美单词

against *prep.* 反对，对阵

junk *n.* 垃圾，废弃物

tend *v.* 倾向于，趋向于

balance *v.* 使平衡，使均衡

sort *n.* 品种，类别

lasagna *n.* 烤宽面条

develop *v.* 发展，壮大

fierce *adj.* 猛烈的，激烈的

advertisement *n.* 广告，宣传

roadside *n.* 路边，路旁

Conversation 2

Welcome! Can I help you?	欢迎光临！请问你需要点什么？
I want a small order of French fries and a Big Mac.	我要一小份炸薯条和一个巨无霸。
Anything else? What about a strawberry pie?	您还需要其他的吗？来一个草莓派怎么样？

No, thanks.	不要了，谢谢你。
Is that for here or to go?	您是在这里吃还是要带走？
For here.	在这里吃。
What would you like to drink, Coca Cola or fruit juice?	您想要点什么喝的，可乐还是果汁？
I'd like an orange juice.	我要一杯橙汁。
Is that all?	就这些吗？
Yes, that's it.	是的，就这些。
Could you wait just a moment, please? Your order will be ready soon.	您能稍等片刻吗？你要的东西很快就好。
Sure!	当然可以。

开胃词组

to go 外卖，外送

French fries 炸薯条

Big Mac 巨无霸

that's it 够了，好了，就这样吧

wait a moment 稍等一会儿

鲜美单词

strawberry *n.* 草莓

French fries *n.* 炸薯条

Coca Cola *n.* 可口可乐

order *n.* 点菜，订餐

drink *n.* 酒，饮料

juice *n.* 果汁

wait *v.* 等候，等待

ready *adj.* 准备好的

Origins of Fast Food in the USA 美国快餐史

The USA is usually thought of as the fast food capital of the world, and blamed for all sorts of junk food related ills and assaults on the global diet. How did the land most associated with burgers and hot dogs come to be such a massive fast food haven?

美国常常被称为快餐王国，也因为其垃圾食品给世界饮食带来了不好的冲击而备受指责。那么，汉堡和热狗是如何造就了这样的快餐帝国呢？

The Hot Dog – the USA's Original Fast Food?

热狗——美国最初的快餐

In 1867, a German butcher called Charles Feltman opened the first hot dog stand in Brooklyn, New York. The delicacy caught on with the native New Yorkers and a USA favourite food was born. The spread of the hot dog is said to be attributed to the World's Columbian Exposition in 1893. Also, in Chicago, the St. Louis World's Fair, 1904, where a number of new fast foods were promoted as desirable to the general public, including hot dogs and ice cream cones.

早在1867年，一位名叫查尔斯·费特曼的德国肉铺老板在纽约布鲁克林区开了第一个热狗摊。它很快成为纽约人喜欢的食物，美国人最喜欢的食物就此诞生了。热狗的流行也得益于1893年的哥伦比亚世界博览会。然后1904年在芝加哥的圣路易斯世界博览会上，大量快餐涌入市场，包括热狗和冰淇淋蛋卷。

Burgers and Fries

汉堡和薯条

The origins of fast food restaurants in the USA can probably be traced to a specific date - 7 July, 1912 – when a fast food restaurant was opened in New York City by Horn & Hardart. The establishment offered its happy customers a selection of pre-prepared fast foods which were displayed behind small glass windows and coin-operated slots.

美国最早的快餐店可追溯到1912年7月7日霍恩和哈德特在纽约创办的快餐店。他们把快餐陈列在玻璃窗后面，使顾客愉快地挑选所喜爱的套餐并使用投币的方式购买。

Fast Food in the 21st Century

21世纪的快餐行业

Fast food makes up a large proportion of the staple diet of some Americans. In the USA alone, it was estimated that around US$110 billion was spent just on fast food in 2000 and according to the National Restaurant Association, sales of fast food in the USA were up to $163.5 billion in 2005.

快餐是美国人的日常食品。美国每年快餐销售额，2000年已经达到1100亿美元。根据国家餐饮协会数据显示，2005年提高到了1635亿美元。

Part 4

British Cuisine
英伦风味

National Dishes
经典国菜

舌尖美食词汇

fish and chips 炸鱼和薯条	**chicken tikka masala** 咖喱鸡块
Scotch broth 苏格兰浓汤	**Welsh cawl** 威尔士炖菜
stovies 苏格兰土豆炖菜	**Black Bun** 苏格兰黑面包
haggis 肉馅羊肚	**Yorkshire pudding** 约克郡布丁
Cornish pasty 康瓦尔郡菜肉煤饼	**Glamorgan sausage** 格拉摩根香肠

舌尖美食句

1. Chicken tikka masala is thought to be the national dish of England.	咖喱鸡块被认为是英国的国菜。
2. I just ordered Scottish beef and Scotch broth.	我刚点了苏格兰牛肉和苏格兰浓汤。
3. Here is the recipe for preparing fish and chips.	这是炸鱼和薯条的菜谱。

	English	中文
4.	England's national dish is said to be fish and chips.	据说炸鱼和薯条是英国的国菜。
5.	It is said fish and Chips is a real British treat.	炸鱼和薯条是英国真正的美食。
6.	Could you recommend some typical food to me?	你可以推荐几个特色菜吗？
7.	Chicken tikka masala is a delicious dish.	咖喱鸡块是一道美味的菜。
8.	Stovies is a Scottish dish based on potatoes.	苏格兰土豆炖菜是一道以土豆为主料的苏格兰菜。
9.	Black Bun is a rich and delicious fruit cake served at Hogmanay.	苏格兰黑面包是苏格兰除夕时吃的美味水果糕点。
10.	Haggis is a must on the menus.	肉馅羊肚是一道必点的菜。

舌尖聊美食

Conversation 1

Hi, Lily. Did you travel to England last week?	嗨，莉莉。你上周去英国旅游了吗？
Yes, I did. I toured around a couple of days there. I had a lot of fun.	是的，我在那游玩了几天。非常开心。
Did you eat some special native food?	你吃了当地的美食了吗？
Well, yes. Fish and chips are very impressing. That's a classic national dish.	是的。炸鱼和薯条令人印象深刻。那可是一道经典国菜。

It's said that fish and chips is a must on any British menu. Do you know how to make it?

据说炸鱼和薯条是在英国必点的一道菜。你知道怎么做这道菜吗？

It's quite easy. First, you prepare the ingredients, including potato strips, all-purpose flour, baking powder and ground pepper. Then, mix the potato strips together with flour, baking powder, salt and pepper. Stir in the milk and egg until the mixture is smooth. Fry the potatoes in the hot oil until they are tender. Drain them on paper towels. Dredge the fish in the batter, and place them in the hot oil. Fry until the fish is golden brown and take it out. Finally, place the fish and chips in a tray.

相当简单。首先，准备好材料，包括土豆条，通用面粉，发酵粉，还有辣椒粉。然后把土豆条和面粉、发酵粉、盐和辣椒粉等混合在一起。放到牛奶和鸡蛋里拌匀。把它们放到油里炸到软为止。再把它们放到滤纸上。然后，把鱼肉放入面粉，在油里炸至鱼肉变金黄色时就可以取出来，最后将鱼和薯条放到一个盘子里。

Wow, it does sound not so hard. What about the dressing or dipping?

哇，听起来并不难。调味品或蘸酱什么的呢？

Serving with malt vinegar, lemon, or tartar sauce would be perfect.

蘸着麦芽醋，柠檬汁或塔塔酱吃就更完美了。

I see. Bye, then.

明白了。再见了。

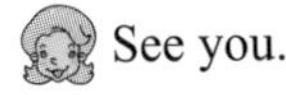

See you.

再见。

开胃词组

a couple of 两个，几个

national dish 国菜，国家菜肴

baking powder 发酵粉

paper towel 纸巾

take out 取出

serve with 向……提供，配以

鲜美单词

native adj. 本地的，当地的

impressing v. 给……以深刻印象

classic adj. 典型的，经典的

strip n. 条，带

stir v. 搅拌，搅动

mixture n. 混合物，混合料

drain v. 滤去，滤干

dredge v. （用面粉，糖等）撒

place v. 放置，把……放在

dressing n. 调料，调味品

malt n. 麦芽，麦乳精

vinegar n. 醋

Conversation 2

Hello. I like to try some authentic Welsh cuisine. Could you recommend some typical ones to me?	您好。我想尝一些地道的威尔士菜。你可以推荐几个特色菜吗?
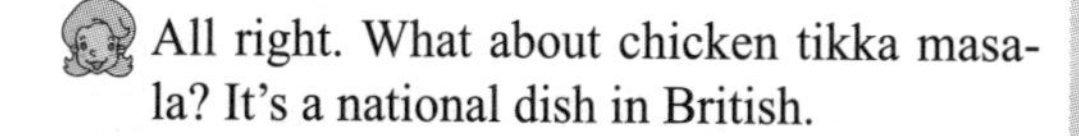 All right. What about chicken tikka masala? It's a national dish in British.	好的。咖喱鸡块怎么样？它可是英国的国菜呢。
 Is it spicy?	它是辣的吗?
 Yes. A little spicy.	是的，有点辣。
 Is the meat tender?	肉嫩吗?
 The chicken is wonderfully tender due to being marinated in yoghurt and Tandoori spices before being baked and then cooked again in a rich Tikka Masala sauce.	它里面的鸡肉非常嫩，因为鸡肉是经过在酸奶和唐杜辣味调料里腌制后，烧烤，然后在咖喱里煮。

Hmm, it serves with bread, doesn't it?	嗯，是和面包一起吃吗?
Yes, toast bread will be fine with it. You also could order a side dish raita to go with it. Raita is a salad consisting of cubed cucumber, onion and tomato in yoghurt.	是的，和烤面包一起吃很不错。酸奶沙拉做配菜也不错的。酸奶沙拉是由黄瓜块、洋葱和土豆在酸奶里拌成的。
OK. I take a chicken tikka masala, one toast bread and a dish of raita.	好的。我点一份咖喱鸡块、一个烤面包和一盘酸奶沙拉。
And for the drink, red wine or champagne will be good.	关于饮料，红酒或香槟酒都不错。
A red wine, please. That's all. Thank you.	来一瓶红酒吧，就这些了。谢谢。
You're welcome. I'll be right back.	不客气。我马上回来。

开胃词组

Welsh cuisine 威尔士菜

Tikka Masala 咖喱

toast bread 烤面包

side dish 配菜

consist of 由……组成

鲜美单词

authentic *adj.* 真正的，正宗的

typical *adj.* 典型的，具有代表性的

spicy *adj.* 辣的，辛辣的

marinate *v.* 浸泡，腌泡

yoghurt *n.* 酸奶

bake *v.* 烤，烘焙

raita *n.* 黄瓜优酪乳，酸奶沙拉

consist *v.* 由……组成， 由……构成

cube *v.* 把（食物）切成小方块

champagne *n.* 香槟酒

舌尖美食文化

Fish and chips is a hot dish of English origin, consisting of battered fish, commonly Atlantic cod or haddock, and deep-fried chips. It is a common take-away food. A common side dish is mushy peas.

炸鱼薯条是一道英国热门菜，里面含有面粉炸鱼，通常是北冰洋鳕鱼或黑线鳕鱼肉和油炸薯条。它也可以打包带走，最常见的配菜是豆泥。

Fish and chips became a stock meal among the working classes in the United Kingdom as a consequence of the rapid development of trawl fishing in the North Sea, and the development of railways which connected the ports to major industrial cities during the second half of the 19th century, which meant that fresh fish could be rapidly transported to the heavily populated areas. Deep-fried fish was first introduced into Britain during the 16th century by Jewish refugees from Portugal and Spain, and is derived from pescado frito. In 1860, the first fish and chip shop was opened in London by Joseph Malin.

炸鱼薯条之所以成为英国普通工人家庭的一道家常菜，是由于19世纪中后期，随着北冰洋渔业的发展和英国铁路的迅猛发展，各大港

口连接了各主要工业城市，从而使大量鲜鱼得以快速地运往人口密集城市。炸鱼烹饪是16世纪由葡萄牙和西班牙犹太难民传入英国的，这种烹饪源自西班牙的炸鱼烹饪法。首家炸鱼薯条店，由约瑟夫·马林于1860年在伦敦创立。

读书笔记

UNIT 02 Traditional Specialties 传统特色菜

舌尖美食词汇

black pudding 黑布丁（猪血香肠）	**Cullen skink** 鲜鱼浓汤
Scottish smoked haddock 苏格兰烟熏黑线鳕	**bubble and squeak** 油煎土豆卷心菜
parsley liquor 西芹汁	**Yorkshire Hot Pot** 约克郡火锅
cottage pie 农舍派	**bangers and mash** 香肠土豆泥
clotted cream 凝脂奶油	**baked beans** 烤豆

舌尖美食句

1. Is there any traditional specialty in Britain? 英国有特色菜吗？
2. Black pudding is the most favorite specialty in England. 黑布丁是英国最受欢迎的特色美食。
3. Cullen skink is a traditional thick Scottish soup originating from the North East of Scotland. 鲜鱼浓汤是起源于苏格兰东北部的传统苏格兰式浓汤。

4. Scottish smoked haddock is the best seller in this restaurant.	苏格兰烟熏黑线鳕是这家饭馆的卖得最好的菜。
5. Why don't we get a bubble and squeak?	为什么我们不来份油煎土豆卷心菜?
6. Parsley liquor is extremely popular in England and Ireland.	西芹汁在英国和爱尔兰非常流行。
7. Steamed steak and onion pudding are lovely for a lunch.	清蒸牛排和洋葱布丁可是午餐美味之选。
8. Some British dishes have very interesting name, such as toad-in-hole.	一些英国菜的名字很有趣，比如“井底之蛙”（香肠吐司）。
9. Yorkshire Hot Pot is a must in this restaurant.	约克郡火锅是这家饭馆的招牌菜。
10. A traditional cottage pie is made with ground beef and topped with mashed potatoes.	传统农舍派是用牛肉肉末做的，上面浇一层土豆泥。
11. Would you like bangers and mash as well?	你也喜欢吃香肠土豆泥吗?
12. Cornwall is famous for its clotted cream.	康沃尔郡因凝脂奶油非常有名。

舌尖聊美食

Conversation 1

Good morning. Would you like to have breakfast with me?	早上好，要不要跟我一起去吃早餐?
Why not? Traditionally, what do English people eat for breakfast?	为何不呢？英国人早餐都吃什么呢?

Well, a full English breakfast contains bacon, sausages, fried tomatoes, fried eggs and toast. Of course, you can replace fried eggs with scramble eggs.	嗯，全套英式早餐包括培根、香肠、炒西红柿、煎蛋还有吐司。当然，你也可以把煎蛋换成炒鸡蛋。
Is it all the same things in Britain?	全英国人都吃这些一样的东西吗？
Not exactly. Some may include fruits, yogurt, whole wheat bread, oatmeal and black pudding. It depends on the personal choice. What would you like?	不一定啊，有的还吃水果、酸奶、全麦面包、燕麦粥，还有黑布丁。那得根据个人喜好了。你想吃什么？
Oh, I would like a black pudding, some fruits and a bowl of oatmeal.	噢，我来一个黑布丁、一些水果和一碗燕麦粥。
For me, I want a sausage, fried eggs and toast, the classic set meal.	我要一根香肠、一个煎蛋和吐司就好了，经典套餐。

开胃词组

fried eggs 煎蛋

scramble eggs 炒蛋

whole wheat bread 全麦面包

depend on 依赖，依靠；随……而定

set meal 套餐

鲜美单词

breakfast *n.* 早餐，早饭

bacon *n.* 培根，熏猪肉

sausage *n.* 香肠，腊肠

replace *v.* 替换，代替

scramble *v.* 炒（蛋）

Britain *n.* 英国

pudding *n.* 布丁

oatmeal *n.* 燕麦粥

Conversation 2

It's becoming more crowded, isn't it?	人越来越多了，对吗？
Yes, people flood in this restaurant for dinner on weekend. It's normal that you may wait for half an hour before you can get a seat.	是的，人们在周末如潮水般涌来这里吃饭。排队半个钟头才能得到一个座位是常有的事。
Wow, unbelievable.	哇，真难以置信。
We are extremely fortunate to come earlier. Umm, I think it's time for us to order.	应该庆幸我们提前来了。嗯，我觉得我们该点菜了。
Good. Peter, is this pudding? I thought it's a sausage.	好，皮特，这是布丁吗？我觉得它好像香肠啊。
Oh, yes, a kind of traditional specialties. Black pudding is a dark sausage filled with meat, barley, oats and other things like that.	噢，对呀，这是一种传统的特色美食。黑布丁是一种黑色香肠，里面填有肉、大麦、燕麦之类的东西，
But why is it called pudding?	不过，它为什么叫布丁呢？
It might be the black sausage which looks like pudding on the first sight.	或许是因为黑色的香肠一眼看上去很像布丁吧。
Oh, I see. I'll try this.	哦，我懂啦。我要尝尝这个。

开胃词组

on weekend 在周末

wait for 等待

fill with 填充，填满

on the first sight 刚看见的时候

鲜美单词

crowded *adj.* 拥挤的，人多的

normal *adj.* 正常的，平常的

unbelievable *adj.* 不可信的，难以置信的

fortunate *adj.* 侥幸的，幸运的

barley *n.* 大麦

oat *n.* 燕麦，燕麦粥

call *v.* 称呼，命名为

sight *n.* 看见；视野

舌尖美食文化

Black pudding, otherwise known as blood sausage, is a dark sausage stuffed with animal blood seasoned and cooked with fillers such as bits of meat, suet, oats, or barley and congealed until solid. Although this dish is normally made with cow or pig blood, it can also be made with the blood of ducks, geese and lambs. The fillers, seasonings, and type of animal blood used vary according to regional tastes and local availability.

黑布丁，也叫黑香肠，是一种用动物血做佐料，并与肉、板油、燕麦、大麦等混合物填充入内的黑色香肠。尽管这道菜一般用牛血或

猪血做原料，但是鸭血、鹅血和羊血也是可以的。根据地方的口味和食材来源的不同，用于制作黑布丁的填充物和佐料就有所不同。

Black pudding is extremely popular in England and Ireland. In northern England, it even has a festival dedicated in its honor: the World Black Pudding Throwing Championships, wherein participants sling black puddings in an effort to knock Yorkshire puddings off a stack. British black pudding is generally made with pig's blood mixed with pork fat and oatmeal or barley. It is traditionally served as part of a full breakfast, but it has become popular as a fried item in fish-and-chip shops.

黑布丁在英国和爱尔兰很受欢迎。在英国北部地区，还有一个有关黑布丁的节日：世界黑布丁大赛。该节日活动就是在投掷车里放黑布丁，投射出去，击中并打掉草堆上的约克郡布丁。英式黑布丁是用猪血混合猪油、燕麦和大麦制作而成的。传统上，它是早餐套餐必备的美味，但是如今在炸鱼薯条店里也常看到油炸黑布丁。

读书笔记

UNIT 03 Fine Main Courses 精致主菜

舌尖美食词汇

main course 主菜	**cream and chive potato** 奶油香葱土豆
braised lamb shank 红焖羊肘	**kedgeree** 鱼蛋饭
lamb curry 咖喱羊肉	**braised fish in brown sauce** 红烧鱼
flapjack 煎饼	**sweetcorn soup** 玉米羹
British whiskey chicken 英式威士忌醉鸡	**Sunday roast** 周日烤肉大餐

舌尖美食句

1. The main course was typically British. 主菜是地道的英国菜。

2. A home-made cream and chive potato could be the most tasty gourmet food in the world. 一份自制奶油香葱土豆可是世界上最美味的珍品。

3. Braised lamb shank is a popular recipe in British cuisine. 红焖羊肘是一道在英国广受欢迎的菜。

4. Would you like a meal deal or just kedgeree alone?	你想吃套餐呢，还是只需要奶油鱼蛋饭？
5. I think you forgot one of our main courses — lamb curry.	我觉得你忘了给我们上那道主菜—咖喱羊肉。
6. You could do a Braised fish in brown sauce as the main course.	你可以点一份美味红烧鱼作为主菜。
7. We have great salad recipes in our appetizer section, like warm balsamic kale salad.	我们有很棒的沙拉作为开胃菜，比如香油甘蓝沙拉。
8. In British food, a flapjack is a sweet bar made from a mixture of oats, corn syrup and butter.	英国的燕麦饼是一种用燕麦、玉米浆和黄油制作的甜饼。
9. Roast lamb is loved for their natural tenderness and sweet flavour.	烤羊肉因其肉嫩味美而广受人们喜爱。
10. The creamy sweetcorn soup is perfect for a weekend lunch.	奶香味的玉米羹是周末午餐的完美汤品。
11. A British whiskey chicken can be served in different restaurants.	英式威士忌醉鸡在不同的饭馆都可以吃到。

舌尖聊美食

Conversation 1

 May I take your order?	你现在要点菜吗？
 Yes, please. I would like a Scotch broth for the starter. Actually, I'm not familiar with British cuisine. Do you have any recommendation for the main course?	是的。我点一个苏格兰浓汤作为开胃菜。实际上，我对英国菜不太了解。主菜方面，有什么推荐吗？

Well, lamb is quite popular in England. How about lamb curry or braised lamb shank?	羊肉在英国相当受欢迎的。来一份咖喱羊肉或红焖羊肘，怎么样？
Oh, I'm sorry I don't eat lamb and pork.	噢，对不起，我不吃羊肉和猪肉。
In that case, I recommend a dish of beef. Scottish beef must be the best choice for you and it's the best seller in our restaurant.	那样的话，我推荐你来一份牛肉。苏格兰牛肉肯定是你的最佳之选，它可是我们饭店的招牌菜。
What part of the beef is it?	选自哪个部位的牛肉？
It's from the ribs. Our chef seeks out well-marbled beef for the most tender and succulent stew. This stew is so intensely flavorful and deserves a rich and powerful red from Bordeaux to go along with it.	肋排。我们的厨师精选细纹牛肉烹制鲜嫩多汁的炖肉。这份炖肉相当别有风味，配上一瓶醇厚的波尔多葡萄酒可再绝妙不过了。
It sounds amazing. I will take one. Thank you very much.	听起来相当棒啊。我来一份吧。谢谢。
You're welcome. And let's move on the others.	不客气！继续点其他的吧。

开胃词组

take order 下菜单，订餐

be familiar with 熟悉，认识

have recommendation for 为……推荐

in that case 如果那样的话

part of 部分

seek out 找到，找出

鲜美单词

broth *n.* 肉汤

starter *n.* 开胃小吃

curry *n.* 咖喱

braise *v.* 炖，闷

shank *n.* 胫，小腿

restaurant *n.* 饭店，餐馆

succulent *adj.* 肉质的，多汁的

stew *v.* 炖，煨

deserve *v.* 应得，值得

Conversation 2

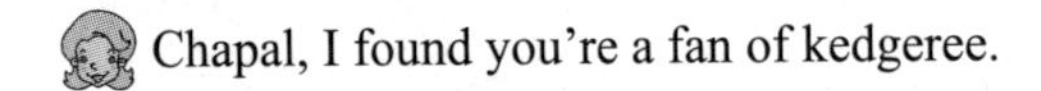

Chapal, I found you're a fan of kedgeree.	查普尔，我发现你真的非常喜欢吃奶油鱼蛋饭。
Oh, yes. That's my favorite main course. Kedgeree helps me to keep in mind the wonderful time during the childhood.	噢，是的。那是我最喜爱的主菜。奶油鱼蛋饭总让我想起童年的美好时光。
Well, I thought kedgeree is a traditional British cuisine.	哦，我还以为奶油鱼蛋饭是一道传统英国菜呢。
I have no idea. But they contain sort of similar ingredients and the like. Probably it was introduced to UK from India.	我不太清楚，但是它们的食材都有点类似。它也许是从印度带到英国的吧。
I agree. Both routes could be possible. But anyway, this dish is widely welcome in both regions.	我同意你的说法。两种情况都有可能。但是，不管怎么说，这道菜在两个地区都很受欢迎。
Great! Let's enjoy it!	太好了！咱们来享用这道美味吧！

开胃词组

keep in mind 记住

have no idea 不清楚，不了解

sort of 有点，有几分地

the like 类似的东西

introduce to 向……介绍

鲜美单词

fan *n.* 迷，粉丝

kedgeree *n.* 奶油鱼蛋饭

wonderful *adj.* 极好的，美妙的

childhood *n.* 童年，童年时期

cuisine *n.* 烹饪，菜肴

contain *v.* 包含，含有

introduce *v.* 介绍，引进

colonist *n.* 殖民者

route *n.* 渠道，途径

region *n.* 地区，地域

舌尖美食文化

Kedgeree is a dish consisting of cooked, flaked fish, boiled rice, parsley, hard-boiled eggs, curry powder, butter or cream and occasionally sultanas.

奶油鱼蛋饭是用鱼片、米饭、香芹、水煮鸡蛋、咖喱、黄油或奶油做成的，有时加点葡萄干。

Kedgeree is thought to have originated with an Indian rice-and-bean or rice-and-lentil dish Khichri, traced back to 1340 or earlier. It is widely believed that the dish was brought to the United Kingdom in Victorian times, part of the then fashionable Anglo-Indian cuisine. It is one of many breakfast dishes that, in the days before refrigeration, converted yesterday's leftovers into hearty and appealing breakfast dishes, of which bubble and squeak is probably the best known.

据说奶油鱼蛋饭来自于1340年的印度或更早些时候的一种印度豆饭或一种小扁豆菜饭。人们认为奶油鱼蛋饭是维多利亚女王时代从印度带回英国的，属于当时印度的流行菜肴。来到英国，这道菜成为早餐特色之一。在电冰箱问世之前，人们把前夜剩饭做成第二天的美味早餐。在英国的早餐中，炸土豆卷心菜最为著名。

读书笔记

Leisure Teatime 休闲茶点

舌尖美食词汇

cake 蛋糕	**sandwich** 三明治
Earl Grey tea 格雷伯爵茶	**coffee** 咖啡
milk tea 奶茶	**cucumber sandwiche** 黄瓜茶点三明治
biscuit 饼干	**buttered toast** 黄油吐司
crumpet 松脆饼	**assorted pastries** 杂饼

舌尖美食句

1. Do most British people have teatime?	大多数英国人都喝下午茶吗?
2. Sometimes people take afternoon tea and have small cakes and sandwiches with their tea.	有时人们喝下午茶的时候,一边喝茶一边品尝小蛋糕或三明治。
3. Do you have Earl Grey at teatime?	你在下午茶时间有格雷伯爵茶吗?
4. This kind of sandwich is too creamy.	这种三明治奶油太多了。

5. Tea time in Britain means more than drinking a cup of tea. Teatime is an actual meal.	英国的下午茶不仅仅是喝茶，下午茶实际上是一顿餐点。
6. Would you like a cup of coffee or tea?	你想来一杯咖啡还是茶?
7. I have an addition to milk tea.	我特喜欢喝奶茶。
8. Cucumber sandwiches are most often served at afternoon tea.	黄瓜茶点三明治是下午茶的茶点。
9. In most British homes, afternoon tea consists of a cup of tea and a light snack, a biscuit, a piece of cake, hot buttered toast or crumpets.	在英国普通家庭里，一般下午茶含有一杯茶水和小加餐、饼干、一块蛋糕、热的黄油吐司或松脆饼。
10. Try our delicious scones for afternoon tea.	喝下午茶，来尝一尝我们美味的司康饼。
11. I recommend assorted pastries for the teatime.	关于下午茶，我建议来点杂饼。

舌尖聊美食

Conversation 1

 Kevin, teatime! Come and get your serving.	凯文，下午茶时间到了！快来拿你的餐点。
 Coming soon, mom!	马上就来，妈妈！
 Look, how splendid they are. What would you like today? I've prepared cucumber tea sandwich, tuna salad tea sandwich, Earl Grey, Darjeeling, muffin and scones with slotted cream and jam.	看，这些餐点多棒！你今天要什么？我准备了黄瓜茶点三明治、金枪鱼沙拉茶点三明治、伯爵茶、大吉岭茶、小松饼和带奶油及果酱的司康饼。

Hmm! Cucumber tea sandwich!	嗯，黄瓜茶点三明治吧！
Hah, you don't like to try new things like tuna salad one?	哈哈，你不尝一下别的新东西吗，比如金枪鱼沙拉三明治？
Cucumber tea sandwich is just my forever love, you know.	你知道，黄瓜茶点三明治是我的最爱。
Do you have some tea?	你要茶水吗？
No tea, thanks, mom. I would like a cup of water then.	不了，谢谢妈妈。我想来杯水就行。
Fine. Take these scones and Darjeeling to your father. He's at the front porch.	好，把这些司康饼和大吉岭茶给你爸爸送过去。他在前面的门廊那里。
OK.	好的。

开胃词组

forever love 永恒之爱，钟爱

coming soon 快到了，马上到

a cup of 一杯

front porch 前廊，走廊外面

鲜美单词

teatime *n.* （英国下午茶的）喝茶时间

splendid *adj.* 极好的，极妙的

cucumber *n.* 黄瓜

Darjeeling *n.* （印度的）大吉岭茶

slot *v.* 放入，插入

jam *n.* 果酱

tuna *n.* 金枪鱼

scone *n.* 烤饼，司康饼

porch *n.* 门廊，走廊

Conversation 2

Coming! Hi, Mrs. Donald. What's up?	马上就来啦！嗨，唐纳德太太！什么事啊？
Hi, Ms. King. I'm sorry to bother you. I just ran out of salt and vinegar. Could I borrow some from you?	你好，金太太！对不起，打扰了。我家的盐和醋用完了，可不可以向你借一点？
No problem. But I am making some pastries for the teatime. Could you come in and wait a second?	没问题。但我正在做下午茶的糕点呢。你能不能进来稍等一下呢？
Sure. I've already smelt it. Yummy!	当然。我闻到香味了。很好闻哦！
Look, that's the donuts, almond croissant and mini muffins. You can take some home to have a taste if you like.	看，那是甜甜圈、杏仁牛角面包和迷你松饼。如果你喜欢，带一点回家尝尝。
Wow, it's really nice of you. Thank you so much. Next time, drop by my house for a drink.	哇，你真是太好了！非常感谢！下次来我家，我们喝点东西。
OK, sure.	好的，当然啦。

开胃词组

run out of 用完，耗尽

no problem 没问题

wait a second 稍等片刻

have a taste 尝一尝

drop by 顺便拜访

鲜美单词

bother *v.* 烦扰，打扰

salt *n.* 盐，食盐

vinegar *n.* 醋，食醋

pastries *n.* 糕点，油酥糕点

donut *n.* 油炸圈饼，甜面圈

almond *n.* 杏仁

croissant *n.* 羊角面包

muffin *n.* 小松饼

taste *n.* 尝，品尝

舌尖美食文化

Tea time in Britain means more than drinking a cup of tea. Teatime is an actual meal and depending upon where in the country you find yourself, this meal could be anything from a scone to a few light sandwiches and cakes, to a full roast dinner.

在英国，下午茶时间不仅仅是喝茶，而实际上是吃餐点的时间。根据不同地方，这个餐点有所不同，有司康饼、三明治、蛋糕，还有烧烤。

In addition to the slices of bread and butter, one would offer thin cucumber sandwiches, or salmon sandwiches as well as cakes, pastries, scones with cream, and of course a Victoria sponge cake. To be able to bake a light and moist Victoria sponge was the true test of one's cook.

除了面包和黄油，下午茶时间，人们通常吃黄瓜三明治、三文鱼三明治、蛋糕、油酥糕点、奶油司康饼，当然，还有维多利亚海绵蛋糕。是否会做松软湿润的维多利亚海绵蛋糕是衡量厨艺水平的一项测试。

The tea, Indian or Chinese or in many houses, both, would be served in silver tea pots and poured into fine china cups. Etiquette books contained whole chapters on the etiquette of Afternoon Tea and the tea dress was created to be worn at such social gatherings. Afternoon Tea was, it should be noted, a social occasion in which only the upper class participated.

无论印度茶还是中国茶，或者在许多人的家里两者均有，都使用银茶壶或精美瓷杯冲饮。礼仪书里应该有下午茶方面的礼仪，在喝茶的社交场合，穿着也是有考究的。重要的一点，下午茶是上流社会的社交场合。

读书笔记

UNIT

05 British Desserts 英式甜品

舌尖美食词汇

British desserts 英式甜点	**strawberry custurd tart** 草莓蛋挞
trifle 乳脂松糕	**cheesecake** 芝士蛋糕
toffee fudge ice cream 太妃巧克力冰淇淋	**Bakewell pudding** 贝克韦尔布丁
crumble 酥饼	**bread pudding** 面包布丁
mince pie 圣诞派	**Cappuccino creme brulee** 卡布奇诺奶油布丁

舌尖美食句

1. British desserts are one of main characteristics of British cuisine.
英式甜点是英国菜的重要特色之一。

2. There are many dessert options that are British favorites.
英国人喜欢的甜品有很多种。

3. The classic strawberry custurd tart has been a key trend this year.
这款经典的草莓蛋挞成为今年的主流甜品。

4.	Do you have some pre-made trifle in this shop?	你们这儿有可以预订的乳脂松糕吗？
5.	These cheesecakes have all been given the British treatment.	这些芝士蛋糕成为英国人的上等甜品。
6.	I think you will love this sticky toffee fudge ice cream.	我认为你会爱上这种黏的太妃巧克力冰淇淋。
7.	Bakewell pudding is often topped with an egg.	贝克韦尔布丁上面一般有个鸡蛋。
8.	We have a couple of recipes about crumble.	酥饼有好几种做法。
9.	Do you think this haggis is too heavy?	你觉得羊杂碎布丁味道很重吗？
10.	Bread pudding is a traditional British dessert.	面包布丁是一种传统英式甜品。
11.	Could you cut this mince pie into four pieces for me?	你能帮我把这个圣诞派切成四块吗？

舌尖聊美食

Conversation 1

Good morning. What would you like to have?

早上好！你想要买点什么？

To tell you the truth, I'm a tourist here in London. I heard that England is famous for its desserts and has all kinds of desserts to choose. I don't know what to do.

说实话，我是一名伦敦的游客。我听说英国的点心很出名。这里有许多甜点可供选择。我不知道该怎么办。

Well, you're right. We do have quite a lot of desserts. In England, dessert includes cake, pudding, pie, egg tart, biscuit, muffin and ice cream, and so on.	嗯，没错。英国的点心有蛋糕、布丁、派、蛋挞、饼干、小松饼和冰淇淋等。
Puddings are quite popular here, aren't they?	布丁在英国很受欢迎，对吗?
Yes. British puddings can be sweet and light or savory and heavy, for example, cabinet pudding is light and sweet, and haggis and York shire pudding are savory and heavy.	是的，英国的布丁有甜的和清淡的，或是香辣等口味较重的。比如干果布丁就是甜的清淡的，羊杂碎布丁和约克郡布丁就是辣的重口味的。
I would like some authentic and traditional pudding.	我想要一些正宗的，传统风味的布丁。
Why don't you try some haggis? Haggis is a traditional Scottish dish, considered the national dish of Scotland.	你要不来些羊杂碎布丁？羊杂碎布丁是一种传统苏格兰美食，被认为是苏格兰的国菜呢。
Great! Two pounds, please.	好极了！请给我来两磅。
Here you are.	给你。

开胃词组

be famous for 因……而著名

choose from 从……之间选择

for example 例如，比如

here you are 给你

鲜美单词

truth n. 事实，真相

tourist n. 旅行者，游客

famous adj. 著名的，有名的

tart n. 蛋挞

sweet adj. 甜的

savory adj. 好吃的，美味可口的

groaty adj. 燕麦的

cabinet n. 橱柜，储藏柜

authentic adj. 真正的，正宗的

national adj. 国家的，全国的

Conversation 2

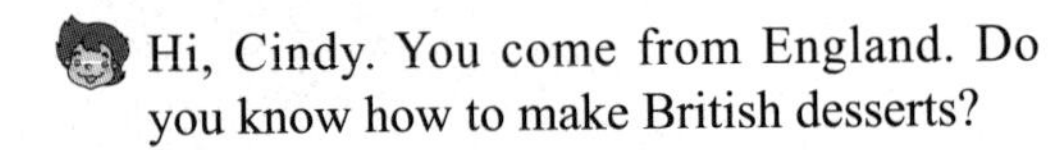

Hi, Cindy. You come from England. Do you know how to make British desserts?

嗨，辛迪！你来自英国，你知道怎么做英式点心吗？

Yes, I do know how to make some desserts. I learned it from my grandma. When I was a child, I used to help out in her dessert shop. I know the recipe about lemon curd, Victoria sponge cake, crumble, trifle and Yorkshire pudding.

是的，我确实知道一些点心的做法。我从我的奶奶那学来的。当我还小的时候，我常常去她的甜品店帮忙。我知道怎么做柠檬奶酪、维多利亚海绵蛋糕、酥皮水果饼、乳脂松糕和约克郡布丁。

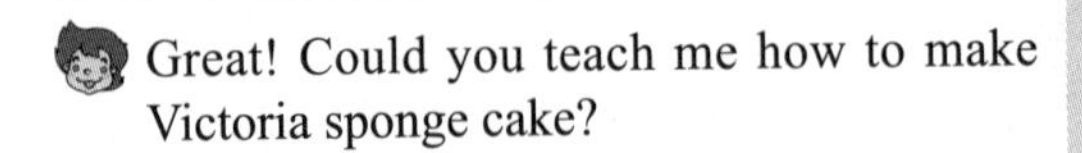

Great! Could you teach me how to make Victoria sponge cake?

棒极了！你能教我做维多利亚海绵蛋糕吗？

Sure. Nothing beats a Victoria sponge cake, and we have loads of recipes for this sandwich cake. I will teach you how to make a soft vanilla sponge cake filled with strawberry, the classic one.

当然。没有什么能比得上维多利亚海绵蛋糕好吃啦。我们有好几种关于它的做法。我教你怎么做香草味的草莓海绵蛋糕，这是其中最经典的一种。

 Wonderful. Come on, go ahead. 好，咱们快点开始吧。

First, you get all the ingredients prepared, including margarine, golden caster sugar, vanilla extract, eggs, flour, butter, icing sugar, strawberry and jam. Then, get two sandwich tins, cream the margarine, caster sugar and vanilla together with a whisk and whisk in the beaten eggs. Sift the flour into the mixture. Split the mixture evenly between the 2 sandwich tins. Bake them until springy and golden brown in an oven. Remove and cool them on a wire rack. Once the sponges are cool, spread the butter cream onto one of the sponges and top with the jam. Place the second sponge on top of the first sponge, round side up, and dust with a little icing sugar. Serve.

首先，你得准备所有的食材，包括人造黄油、金砂糖、香草粉、鸡蛋、面粉、黄油、糖粉、草莓和果酱。然后，准备两个三明治烤盘，用搅拌器把人造黄油、金砂糖和香草粉混合搅匀，打入鸡蛋搅匀，撒入面粉搅拌。把混合面团分别均匀地摊到那两个烤盘。放入烤箱烤到表皮膨起有弹性并呈金黄色，取出，放到过滤网上冷却。海绵蛋糕变凉之后，给其中一块蛋糕涂上黄油奶酪和果酱，再把另一块蛋糕放上去，弄平整，在最上层的蛋糕撒些糖粉。然后就可以上桌了。

 It sounds cool. I will have a try. Thank you. 听起来很棒！我也要试试。谢谢！

 It's my pleasure. 不客气。

开胃词组

come from 来自于，出生在……

help out 帮忙

loads of 大量，许多

fill with 充满，填满

have a try 尝试一下

鲜美单词

recipe *n.* 食谱，菜谱

curd *n.* 凝乳

sponge *n.* 海绵，海绵状物

crumble *n.* 酥皮甜点，酥饼

trifle *n.* 乳脂松糕

beat *v.* 打败，战胜

classic *adj.* 经典的，典范的

margarine *n.* 人造奶油，人造黄油

caster *n.* 调味品瓶

whisk *v.* 搅拌

sift *v.* 撒，筛

split *v.* 分担，分开

icing *n.* 糖衣，糖霜

舌尖美食文化

Cherries jubilee is a dessert dish made with cherries and liqueur (typically Kirschwasser), which is subsequently flambeed, and commonly served as a sauce over vanilla ice cream.

樱桃禧是用樱桃和利口酒（樱桃酒）制作，一种上面浇酒点燃后端出的，配以香草冰淇淋酱的糕点。

The recipe is generally credited to Auguste Escoffier, who prepared the dish for one of Queen Victoria's Jubilee celebrations, though it is unclear whether it was for the Golden Jubilee of 1887 or the Diamond Jubilee in 1897.

这道菜是法国名厨奥古斯特·艾斯可菲的传世之作，是为维多利

亚女王特制的庆典甜品，但是还没确定这道菜是为维多利亚1887年五十周年庆典还是1897年六十周年庆典制作的。

There have been many variations on the idea of flambeed fruit since Escoffier's time, the most famous being Bananas Foster. Other variations include Mangos Diablo (mangos flambeed in tequila) and Pêches Louis (peaches flamed in whiskey).

自从艾斯可菲时代的浇酒甜品诞生以来，已有多种这样的烹制法，最为著名的当属香蕉福斯特火烧冰淇淋。其他的做法有龙舌兰酒烧芒果，还有蟠桃路易斯（威士忌烧蟠桃）。

读书笔记

读书笔记

Part 5

Italian Cuisine
意式美味

UNIT 01

Appetizer 舒心·开胃

舌尖美食词汇

Italian appetizers 意大利开胃菜	**carpaccio salmon** 红鱼子酱鲑鱼片
Bagna cauda 香蒜醍鱼热蘸酱	**Sicilian escarole** 西西里卷心菜
barley mushroom risott 大麦蘑菇烩饭	**Aubergine parmigiana** 帕尔玛干酪茄子
sausage stuffed mushroom 香肠蘑菇	**bruschetta** 意大利普切塔面包片
Parma ham with fresh figs 意大利帕尔马无花果火腿	**Proscuitto and honeydew melon** 风干火腿蜜瓜
sauteed mushrooms 酱汁炒蘑菇	**pesto** 香蒜酱

舌尖美食句

1. The joy of Italian cooking is the simplicity of the dishes.

意式烹饪崇尚简单之美。

2. Selection of Italian appetizers includes cured meats and Mediterranean roast vegetables.

意大利开胃菜精品包括腊肉和地中海烤蔬菜。

3.	I deadly love carpaccio salmon.	我非常喜欢红鱼子酱鲑鱼片。
4.	Bagna cauda has many local variations.	香蒜醍鱼热蘸酱在各地方有不同的风味。
5.	This restaurant is known for its Sicilian escarole.	这家餐馆以西西里卷心菜而闻名于世。
6.	I'm specialized in barley mushroom risotto.	我擅长做大麦蘑菇烩饭。
7.	Aubergine parmigiana is typical Italian starter.	帕尔玛干酪茄子是典型的意大利开胃菜。
8.	The best Italian starter out there is sausage stuffed mushroom.	最好的意大利开胃菜是香肠蘑菇。
9.	I think nothing can beat brilliant bruschetta.	我认为没什么能比得上美味的意大利普切塔面包片。
10.	Parma ham with fresh figs is simple but delicious.	意大利帕尔马无花果火腿简单但很美味。
11.	Proscuitto and honeydew melon is a nice antipasto	风干火腿蜜瓜是很棒的开胃菜。

舌尖聊美食

Conversation 1

Good morning. Thanks for giving me an opportunity for a job interview.

早上好！谢谢您给我这个面试的机会。

We've gone through your resume and found that you have worked in an English restaurant and performed well. But the point is we only operate an Italian cuisine. Though you have an outstanding education background in Italian cooking, I'm still wondering if you're skilled in the regional food of Italy.

我们看了你的简历，发现你以前在一家英国餐馆工作过，表现不错。但是问题是我们是一家经营意大利菜的饭店。尽管你有出色的意大利菜的学习背景，我还想知道你在意大利地方菜方面有没有专长？

I think I am qualified for the Italian cooking, because I'm the only chef in charge of Italian food in that English restaurant. And due to my excellent work, they sent me to Italy for an exchange program. I have learned a lot of traditional Italian cuisine in this program.

我觉得我有资格做意大利菜，因为在那家英国餐馆，我是唯一一位负责意大利菜的厨师。由于我杰出的工作能力，被派往意大利进行交换学习。在这个项目中，我学习了许多传统意大利菜。

Well, that's nice. Are you specialized in starters? As this would be the first impression for the customers.

嗯，很好。你擅长做开胃菜吗？那可是顾客的第一印象。

Yes. I'm good at different kind of starters, such as bocconcini salad, barley mushroom risotto, classic bruschetta, fennel and orange salad…

是的，我擅长各种开胃品，比如意大利干酪球沙拉、大麦蘑菇烩饭、经典普切塔面包、茴香橘子沙拉等等。

Wonderful. Tomorrow we would like to test your practical skills in the kitchen. Good luck with your final interview.

太棒了。明天我们想看看你的厨房实操技能展示。祝你最后的面试好运！

Thanks a lot. I will go all for it.

非常感谢！我会全力以赴的。

开胃词组

thanks for 感谢

go through 翻阅，翻找

in charge of 负责，主管

due to 因为，由于

be specialized in 擅长……方面

鲜美单词

opportunity *n.* 机会，时机
interview *n.* 采访，面试
resume *n.* 简历
Italian *adj.* 意大利的，意大利人的
perform *v.* 表现，运行
outstanding *adj.* 杰出的，显著的
qualify *v.* （使）具有资格，有权
exchange *n.* 交换，互访
traditional *adj.* 传统的，惯例的
specialize *v.* 专门从事，专攻
impression *n.* 印象，感想
starter *n.* 开胃菜
bruschetta *n.* 意大利普切塔烤面包
fennel *n.* 茴香

Conversation 2

Today we're going to do some authentic Italian starters, marinated olives, Aubergine parmigiana, sausage stuffed mushroom, crostini, prosciutto con melone, carpaccio salmone.

今天我们来做一些正宗意大利开胃品，腌橄榄、帕尔玛干酪茄子、香肠蘑菇、开胃面包片、火腿蜜瓜、红鱼子酱鲑鱼片。

OK. I just cleaned up all the kitchenware. What should we do now?

好的。我刚收拾好了厨具。我们现在做什么？

We're ganna start with the carpaccio salmon. You prepare the ingredients, including 350g smoked salmon, half cup of olive oil, 2 tablespoons lemon juice, 1 small red onion, finely chopped, 1 tablespoon parsley, finely chopped, 2 teaspoons capers, ground black pepper and extra capers for garnish.

我们先做红鱼子酱鲑鱼片。你先准备好材料，350克烟熏鲑鱼，半杯橄榄油，2个大汤匙的柠檬汁，1个小红洋葱，切碎，1个大汤匙的西芹，切碎，2个茶匙的刺山柑粉，辣椒粉和一些为配菜准备的刺山柑粉。

I will go and get them. Then what should we do next?

我马上就去准备。然后，我们怎么做?

I will combine oil, lemon juice and capers in a bowl. Mix them well. Then, arrange smoked salmon on individual plates and drizzle dressing over the salmon. And season with ground black pepper and sprinkle with parsley.

我会把橄榄油、柠檬汁和刺山柑粉放到碗里，拌匀。然后把烟熏鲑鱼片单独放到一个盘子里，把调料涂到其他鲑鱼片上面。最后撒一些黑椒粉和西芹。

That's it?

这样就完成啦?

Well, remember to serve immediately, otherwise, the lemon juice will 'cook' the salmon.

嗯，别忘了立即端上桌，要不然柠檬汁就浸入鲑鱼片啦。

OK. Let's rush into it.

好的。我们赶紧做吧。

开胃词组

clean up 打扫，清理

start with 以……开始

lemon juice 柠檬汁

rush into 匆忙……

鲜美单词

marinate v. 浸泡，腌泡

parmigiana adj. 帕尔玛干酪调制的

crostini n. 覆盖着各式配料的小面包

prosciutto n. 意大利熏火腿

carpaccio n. 意式生腌肉片

salmon n. 鲑鱼

kitchenware n. 厨具，炊具

tablespoon n. 大汤匙，大调羹

chop v. 砍，伐，劈

parsley n. 欧芹

caper n. 刺山柑

garnish n. 配菜，装饰菜

arrange v. 排列，布置

drizzle v. 淋，洒

sprinkle v. 洒，撒

舌尖美食文化

Antipasto means “before the meal” and is the traditional first course of a formal Italian meal. Traditional antipasto includes cured meats, olives, peperoncini, mushrooms, anchovies, artichoke hearts, various cheeses (such as provolone or mozzarella), pickled meats and vegetables in oil or vinegar.

意式开胃菜就是餐前点，是意大利菜式传统的第一道菜。传统意式开胃菜包括腊肉、橄榄、意式辣香肠、蘑菇、鳀、洋蓟芯、各式奶酪（如菠萝伏洛干酪、意大利干酪）、腌肉、腌蔬菜等等。

The contents of an antipasto vary greatly according to regional cuisine. It is quite possible to find in the south of Italy different preparations of saltwater fish and traditional southern cured meats, whereas in northern Italy it will contain different kinds of cured meats and mushrooms and probably, especially near lakes, preparations of freshwater fish. The cheeses included also vary significantly between regions and backgrounds.

意大利不同地区有各自的特色开胃品，意大利南部有各种海鲜特色和传统南方风味腌肉，而北部以各式腌肉、蘑菇为特色，尤其湖区的淡水鱼。不同地区干酪也各有特色。

Many compare antipasto to hors d'oeuvre, but antipasto is served at the table and signifies the official beginning of the Italian meal. It may also be referred to as a starter, or an entree.

很多人把开胃菜称为主菜以外的食品，但它是意大利人正式用餐的第一道菜, 即餐前点或头盘。

读书笔记

Italian Pizza 意式比萨

UNIT 02

舌尖美食词汇

pizza al formaggio 奶酪比萨	**pizza with clam** 蛤蜊比萨
Neapolitan pizza 那不勒斯比萨	**pizza Margherita** 玛格丽特比萨
diavola pizza 意式腊肠比萨	**pizza with olive** 橄榄比萨
Pizza Romana 罗马乳酪比萨	**Pizza Viennese** 维也纳比萨
Pesto pizza 香蒜酱比萨	

舌尖美食句

1. Italian pizza is a very popular recipe all over the world.	意大利比萨是世界各地很受欢迎的食谱。
2. The modern pizza was invented in Naples, Italy.	现代比萨是意大利的那不勒斯人发明的。
3. What's the topping on this kind of pizza?	这种比萨上面的配料是什么?
4. Pizza al formaggio is my favorite.	奶酪比萨是我的最爱。
5. How does the pizza with clam taste?	蛤蜊比萨吃起来味道怎么样?

6.	Authentic Neapolitan pizzas are typically made with tomatoes and Mozzarella cheese.	正宗的那不勒斯比萨是以番茄和意大利干酪为配料的。
7.	Could you make pizza sayur?	你会做意式蔬菜比萨吗？
8.	Sicilian pizza is a pizza prepared in thick-crust that originated in Sicily, Italy.	西西里比萨是意大利西西里地区的一种厚底比萨
9.	Pizza vegetarian is amazing for me.	我觉得蔬菜比萨非常棒。
10.	We have to finish this pizza Romana in 10 minutes.	我们在10分钟之内得赶紧吃完这个罗马干酪比萨。
11.	Do they usually put hard boiled egg on pizza Margherita?	他们会把煮好的鸡蛋放到玛格丽特比萨上吗？

舌尖聊美食

Conversation 1

Hi, Helen. I get into some trouble when I make Neapolitan pizza.	嗨，海伦。我做那不勒斯比萨的时候遇到了点麻烦。
OK, what's your trouble?	什么麻烦？
The dough is too wet. How should I do?	我的面团太湿了，该怎么办？
Don't worry. It's not a big deal. You just add a bit more flour to it. It will be fine.	别担心。没什么大不了的，你往里再加一些面粉就好了。
OK. I will try. I also want to add more topping. Could you give me some suggestions?	好的，我试试。我还想在上面添加一点配料，你可以给我一些建议吗？

First, you could add ham on it. Also, spicy sausage, chill, tomato, spinach, cheese are all good choices. Even you could add a whole egg on top.	首先。你可以在上面加点香肠，辣香肠、辣椒、西红柿、菠菜，还有奶酪什么的都是好的选择。甚至整个鸡蛋都可以加上去。
But I don't know which one would be better.	但是我不知道哪种会更好。
Why not do half and half.	为何不尝试各自一半？
Oh, yes! Wonderful! Thanks for your help.	哦，对！好极了！非常感谢你的帮助。
You're welcome.	不客气。

开胃词组

get into trouble 遇到麻烦，陷入困境

not a big deal 没什么大不了的

on top 在上面

half and half 各半，一半一半地

鲜美单词

trouble n. 麻烦，烦恼

pizza n. 比萨

Neapolitan adj. 那不勒斯的，那不勒斯人的

flour n. 面粉

dough n. 生面团

topping n. 浇头，配料

add v. 增加，补充

spicy adj. 辛辣的

spinach n. 菠菜

choice n. 选择，选择权

Conversation 2

Is this the famous al prosciutto?	这就是著名的意大利熏火腿比萨吗?
Yes. That's exactly what we're eating.	是的，那正是我们正在吃的东西。
But I found they are quite different.	但是我发现它们真的不一样。
Yes, they show some differences. Nearly every area of our country has its own type of prosciutto. The two most popular types are prosciutto di Parma and prosciutto di San Daniele.	没错，看起来有点不一样。我们国家几乎每个地区都有自己风味的熏火腿。在意大利最流行的是帕尔马熏火腿和圣丹尼尔熏火腿。
So it seems that this kind of pizza shows regional variations.	这么说，这款比萨也有地域差异啊。
You've got it. But the basic flavor remains the same, salty.	没错，不过各地比萨保持一致的风味，那就是咸味的。
That's my favorite.	那是我的最爱!

开胃词组

be different 与……不同

area of ……的区域

most popular 最流行的

it seems that 看起来，似乎

regional variation 区域变化

鲜美单词

famous adj. 著名的，出名的

exactly adv. 确切地，精确地

difference n. 差别，差异

area n. 地区，区域

country n. 国家，国民

type n. 类型，类别

regional adj. 地区的，区域的

variation n. 变化，变动

flavor n. 味，味道

salty adj. 咸的

舌尖美食文化

Italian pizza may be widely known to have invented only the last decades of the 19th century, although some believe that it was the Greek who first made them, when a cook was summoned by his King and Queen to make a local specialty. He then made rounded bread with topping and colored it with Italian flag colors. He used Italian ingredients to make the colors particularly the tomato sauce for red, mozzarella cheese for white and basil leaves for green. The delicacy was named after the Queen Mergherita and was considered as the first Italian pizza. From that time on, Italian pizza has become the most favorite pizza around the world.

虽然有人认为比萨是由古希腊人首创的，但众所周知，意大利比萨是19世纪末创制的。当时的国王与王后邀请一位厨师来烹饪他的地方特色菜。他在圆面团上面洒配料，做成像意大利国旗的颜色。他用番茄酱来做红色部分，马苏里拉奶酪来做白色部分，罗勒叶来做绿色

部分。这种佳肴美味以玛格丽塔王后的名字命名，被传为最初的意大利比萨。从那时起，意大利比萨变成为享誉世界的美食。

Today, making a pizza might be easily made, as many instant crust and topping are already available in the market. However, the true taste of Italian pizza is hard to duplicate. Only the best Italian pizza maker can give the consumer the true taste of a tasteful and delicious Italian pizza.

如今，做一份比萨也许很简单，因为许多即用面食和配料在市场都有出售。然而，纯正口味的意大利比萨是无法复制的。只有最出名的意大利比萨烹饪师才能制作最正宗的意大利比萨。

读书笔记

UNIT 03

Special Main Courses 特色主菜

舌尖美食词汇

grilled beef tenderloin fillets 烤牛肉里脊	**stuffing chicken rolls** 酥皮鸡肉卷
baked cod fish 纸包银鳕鱼	**battered chicken breast** 油炸酥皮鸡胸
grilled lamb chop 香草羊排	**smoked duck breast** 烟熏鸭胸
chicken cacciatore 意式鸡肉蔬菜煲	**chicken pancake with herb** 香草鸡肉饼
veal escalope with mushrooms 蘑菇牛肉片	**minced stockfish salad** 鳕鱼干条沙拉

舌尖美食句

1. The main course is called secondo in Italy. 在意大利，主菜就是第二道菜。
2. Grilled beef tenderloin fillets are one of my favorite cuts of meat. 烤牛肉里脊是我最喜欢吃的肉菜。
3. Stuffing chicken rolls are easy to prepare. 酥皮鸡肉卷很容易做。

4.	We only have some baked cod fish left now.	我们现在只剩纸包银鳕鱼了。
5.	You need to preheat the oil to make battered chicken breast.	你需要先把油预热来做油炸酥皮鸡胸。
6.	How can anyone resist a perfectly grilled lamb chop?	谁能阻挡香草羊排的美味诱惑?
7.	Smoked duck breast with orange sauce makes a great addition to a cured meat.	烟熏鸭胸配甜橙汁与腊肉真是相得益彰的美食搭配。
8.	Chicken cacciatore is a tasty tomato and chicken dish.	意式鸡肉蔬菜煲是一道有美味的西红柿和鸡肉的菜。
9.	A large portion size of chicken pancake with herb could satisfy your appetite.	一大份的香草鸡肉饼能让你大饱口福。
10.	Veal escalope with mushrooms delivers the flavor of fresh vegetables.	蘑菇牛肉片散发着鲜美蔬菜的香味。
11.	I'm a super fan of minced stockfish salad.	我是鳕鱼干条沙拉的超级粉丝。

舌尖聊美食

Conversation 1

	Are you ready to order?	可以点菜了吗?
	Yeah I think so. As in an Italian restaurant, I would like to start with sauteed mushrooms for the antipasto.	是的，可以。在意大利餐厅吃饭，我想点一份酱汁炒蘑菇作为开胃菜。
	Great. And what's for the primo?	好的，头盘菜点什么?

Well, I think it must be the pasta, as usual… Oh, hold on! I would like to try something new. Umm…OK, a minestrone soup, please.	嗯，跟平常一样，肯定点意大利面。噢，等一等！今天想来点新的。嗯……好，一份意大利蔬菜汤吧。
And the secondo, I mean the main course?	那第二道菜呢，我是说主菜。
I have no idea. Do you have any recommendation?	我不知道点什么。你有什么可以推荐的吗？
In Italy, for the secondo, chicken, meat, or fish are the usual choices.	在意大利，主菜一般是鸡肉、肉类或鱼肉。
Options are generally small, aren't they? OK, I will take chicken, please.	选择的余地很小，是吧？好，我来鸡肉的。
How about battered chicken breast? The chicken meat is crispy and tender.	油炸鸡肉脯怎么样？鸡肉又脆又嫩。
Fine. It sounds delicious.	好。听起来很美味。
And what's for the contorno and the dolce to accompany the main course?	配菜和甜点吃什么呢？
Minced stockfish salad and mascarpone cream, please.	来一份鳕鱼干条沙拉和一份马斯卡普尼奶酪。
That's all? OK, I will be right back.	就这些啦？好的，我马上就回来。

开胃词组

be ready to 准备就绪

as usual 跟平常一样

hold on 稍等，等一下

鲜美单词

antipasto *n.* 开胃菜

primo *n.* 第一，首先

pasta *n.* 面团，意大利面食

minestrone *n.* 意大利杂菜汤

recommendation *n.* 推荐

option *n.* 选择，选择权

batter *n.* 糊状，面糊

crispy *adj.* 脆的，酥脆的

dolce *adj.* 甜的，甜美的

accompany *v.* 陪伴，陪同

mince *n.* 肉末

mascarpone *n.* 马斯卡普尼干酪（一种牛奶软干酪）

Conversation 2

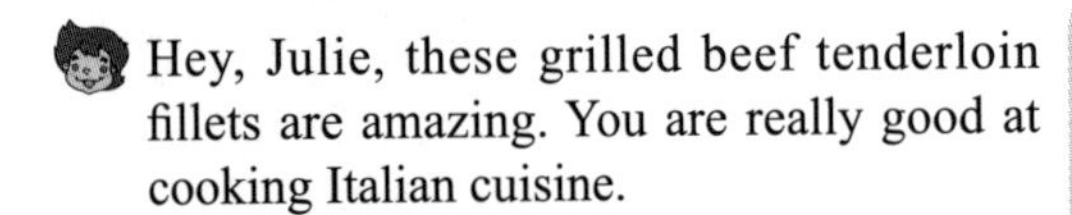

Hey, Julie, these grilled beef tenderloin fillets are amazing. You are really good at cooking Italian cuisine.

嘿，朱莉，这些烤牛肉里脊太好吃了。你做意大利菜非常拿手。

Thank you. My kids are away from home to school, so I spend a little more on the meal and prepare something that tastes enjoyable!

谢谢！我的孩子都去上学了，所以有空做这些好吃的。

Well, could you share the recipe with me?

那你能告诉我怎么做吗?

Why not? Well, for the ingredients, you may need beef tenderloin fillets similar in size, olive oil, dry red wine, garlic minced, Italian herb paste, Kosher salt and fresh ground pepper. Then, place fillets in shallow dish. Mix together balance of the ingredients and pour over fillets, marinate for at least a half an hour. And preheat grill to medium heat. Place fillets on grill and cook 3 minutes on four sides. Time on the grill will vary dependent on the thickness of your cut and the way you prefer to cook the meat. I like mine on the medium well side. You can make mushrooms or garlic for the dressing. That's done!

为什么不呢？食材方面，你需要同样大小的牛肉里脊片、橄榄油、干红葡萄酒、蒜末、意大利香草酱、犹太盐和新鲜胡椒末。然后把里脊牛肉片放在浅盘子里，把配料混合好，倒在牛肉上腌制至少半个小时。然后，把烤架预热到中等程度。把牛肉里脊片放到烤架上每面烤3分钟。还可以根据牛肉的厚度和熟度来决定炙烤的时间。我喜欢中等热度。你可以做些蘑菇或大蒜作为牛肉里脊的调味品。这道菜就这样完成了！

What about the sides?

配菜呢？

You can serve with some amazing sides, like roasted garlic smashed potatoes and lemon butter asparagus!

你可以用精美的配菜，比如烤大蒜土豆或奶油芦笋！

Thank you. I will give it a try soon.

谢谢。我会马上尝试做这道菜。

开胃词组

be away from 远离，离开

similar in 在……方面相似

at least 至少

shallow dish 浅盘

medium well 七八分熟

serve with 向……提供

鲜美单词

tenderloin *n.* 腰部嫩肉，里脊肉

fillet *n.* 肉片

amazing *adj.* 惊人的

enjoyable *adj.* 愉快的，快乐的

herb *n.* 香草

balance *n.* 平衡

preheat *v.* 预热

medium *adj.* 中等的，中级的

vary *v.* 不同，不同于

thickness *n.* 厚度

garlic *n.* 蒜，大蒜

asparagus *n.* 芦笋

Baked cod is a quick and easy dish, and if you don't add a lot of fatty ingredients, it can be healthy as well. The mild taste of cod can be dressed up with a variety of herbs and vegetables, but if you're in a hurry, baked cod can be as simple as drizzling butter and lemon over a thawed filet and sliding it into the oven. Generally speaking, baked cod is ready to eat in about 15 to 20 minutes, though it tends to dry out while cooking, so remember to use a little olive oil, butter or basting liquid to keep it moist.

纸包鳕鱼是制作简单快捷的一道菜，如果你不放很多富含脂肪的原料，就还是一份健康的美食。鳕鱼的清淡芳香可以配上各种香草或蔬菜，如果匆忙的话，给肉片涂上一些黄油或柠檬，放入烤箱即可。一般而言，纸包鳕鱼烤15分钟到20分钟就可食用。烤的时候，鱼肉很容易变干，所以最好放一点橄榄油、黄油或涂油以保持肉质鲜嫩。

UNIT 04 The Great Pasta 意面之缘

舌尖美食词汇

agnolotti 意式方饺子	**rotini** 意式螺旋面
tortiglioni 螺旋通心粉	**conchiglie** 贝壳形意面
farfalloni 蝴蝶形意面	**penne** 斜管面
marinara 海员沙司	**cavatappi** 螺旋扭面
pappardelle 意大利宽面	**lasagna** 意大利千层面
spaghetti 细意大利面	**chitarra** 细条实心面

舌尖美食句

1. Pasta is a very versatile food and can be served in a never-ending array of ways.	意大利面种类繁多，它的配料也是各式各样的。
2. Could you give me a bowl of agnolotti?	你能给我一碗意式方饺子吗？
3. Do you serve rotini in this restaurant?	你们餐馆有意式螺旋面吗？
4. Tortiglioni is my favorite type of pasta.	螺旋通心粉是我最喜欢的意面。

5. The shape of conchiglie looks lovely.
贝壳形意面的外形看起来真可爱。

6. Farfalloni is the classic type of pasta.
蝴蝶形意面是经典的意面。

7. Penne is traditionally served with pasta sauces such as pesto, marinara, or arrabbiata.
斜管面传统上以香蒜沙司、海员沙司和辣味香料番茄酱做调味品。

8. Cavatappi goes well with sauces that are thick or chunky.
螺旋扭面适合搭配厚重的调味酱料。

9. Typically, pappardelle is made with egg-based dough, making the pasta rich in proteins.
通常意大利宽面是鸡蛋面团做的，一种富含蛋白质的意大利面食。

10. Lasagna's classic taste and style will create a memorable evening for your entire family.
意大利千层面的经典风味会给你的家人带来难忘的回忆。

11. Chitarra are long pasta strands, which resemble spaghetti.
细条实心面是长条意大利面，与意大利细面条类似。

舌尖聊美食

Conversation 1

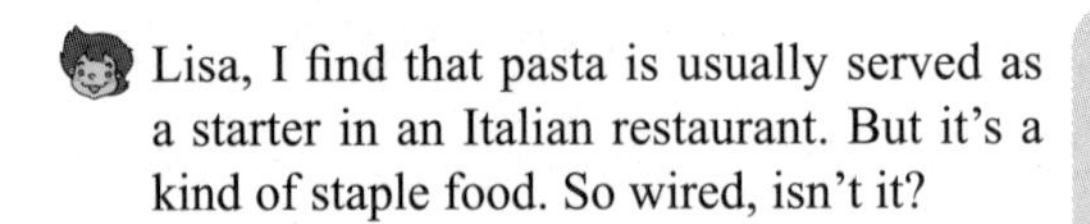

Lisa, I find that pasta is usually served as a starter in an Italian restaurant. But it's a kind of staple food. So wired, isn't it?
丽莎，我发现意大利餐馆把面条当作开胃菜。但是那是一种主食呀，真奇怪。

Well, pasta is a common dish in Italy. It could be regarded as a first course if they are served in small size.
意大利面是意大利一道非常普遍的菜式。小分量的话，可当作开胃菜。

Yes. It could be accompanied with meat for the starter.	是啊，放点肉就可当作开胃菜。
It also can be served as the main dish if it is in a large size.	大分量的话，就可以当作主菜。
Oh, look at this menu, that's the one we had last time. We had tomato sauce for it.	哦，看这菜单，这就是我们上次吃的。我们在里面放了番茄酱。
Yes. This time, why not try pasta in the first course?	没错。我们这次把面条当开胃菜吧，试试看。
Good idea.	好的。

开胃词组

staple food 主食

common dish 常见的菜肴

be regarded as 被认为

first course 第一道菜

last time 上次

鲜美单词

pasta n. 意大利面

staple n. 主食

common adj. 常见的，普遍的

regard v. 认为，看作

accompany v. 伴随

menu n. 菜单，菜肴

sauce n. 调味汁，酱汁

course n. 菜肴，一道菜

Conversation 2

It's time for dinner. I'm starving to death.	该吃晚饭了。我快饿死了。
How about dinning out? I know a new Italian restaurant. They serve authentic Italian noodles.	出去吃怎样？我知道一家新开的意大利餐馆，他们有正宗的意大利面条。
Good choice. But we will be hungry late in the night if we take that only.	不错的选择。如果我们只吃面条，晚上会饿的。
Well, that will not happen to Italian noodles. There're many options among pasta dishes. You'll never walk away from the table hungry.	意大利面不会这样。意大利面有很多种选择的。吃完了绝对不会饿。
Wow, it seems that it is an excellent way to fill the belly.	哇，听起来意面好像很能填饱肚子。
That's right. I promise you will enjoy every single bite of it.	没错，我保证你会非常喜欢吃。
I'm looking forward to it.	我很期待。

开胃词组

it's time for 该是……的时候了

starve to death 饿死

dine out 出去吃饭

late in the night 深夜

look forward to 期待，期望

鲜美单词

dinner n. 正餐，主餐

starve v. 挨饿，饿死

death *n.* 死亡

authentic *adj.* 真正的，正宗的

choice *n.* 选择，选择权

noodle *n.* 面条

excellent *adj.* 卓越的，杰出的

belly *n.* 肚子，腹部

promise *v.* 承诺，保证

bite *n.* 一点食物

Penne is a tube-shaped pasta that originates in Campania, a region in Southern Italy. It is probably one of the more well-known pasta shapes, available in most markets and grocery stores that stock pasta. Cooks with access to a pasta extruder can also make their own penne, should they so desire. Dishes made with it are frequently on the menu at Italian restaurants, especially in the United States, where consumers have a fondness for this shape.

斜管面是源自于意大利南部坎帕尼亚地区的管状的意大利面。这可能是一种最著名的意大利面形状，在超市或杂货店均可买到。有面食挤压机的话，可以自己制作斜管面。意大利餐厅常常有这道菜，尤其在美国，越来越多的顾客喜欢这种意大利面。

The name "penne" comes from the Italian word for "pen," a reference to the angled ends of the tube, which resemble the tip of a quill pen. It comes in smooth and ridged varieties, and can be used in a wide assortment of dishes, from casseroles to soups. The tubes are usually relatively short, around the length and width of a pinkie finger.

斜管面的名字来自于意大利文字“笔管”，它末端的管状与羽毛笔有点相似。它的外皮有光滑和棱纹两种，可做多种菜式，包括炒的或带汤的等等。斜管面长度一般比较短，长宽如小指。

读书笔记

UNIT 05

Desserts 甜品风情

舌尖美食词汇

panettone 潘妮朵尼面包	**pandoro** 潘多洛糖霜面包
cannoli 奶油煎饼卷	**biscotti** 意大利脆饼
panforte 潘福提硬蛋糕	**semifreddo** 冰淇淋蛋糕
zeppole 炸面饼圈	**Neapolitan ice cream** 那不勒斯冰淇淋
Gelato 意大利手工冰淇淋	**bruttiboni** 意大利杏仁酥饼干
tiramisu 提拉米苏	**coconut pie** 椰丝派

舌尖美食句

1. Milano's panettone is the most famous bread of the Italian Christmas cakes.
米兰的潘妮朵尼面包是最负盛名的意大利圣诞水果面包。

2. Pandoro has become very popular around the world.
潘多洛糖霜面包风靡全世界。

3. Cannoli consists of tube-shaped shells of fried pastry dough.
奶油煎饼卷外层包着管状面包皮。

4. If you want to show someone your care, send a box of authentic biscotti.	如果向别人表示关心，就送一盒正宗的意大利脆饼。
5. Panforte is a traditional Italian dessert, containing fruits and nuts.	潘福提硬蛋糕是传统的意大利甜品，内含水果和坚果仁。
6. You can buy a semifreddo at a stand in the street.	在街上小摊，你可买到冰淇淋蛋糕。
7. Zeppole tastes sweet and flavorful.	炸面饼圈香甜可口。
8. Would you like a slice of Neapolitan ice cream?	你想来点那不勒斯冰淇淋吗?
9. Gelato is made with milk, cream, various sugars.	意大利手工冰淇淋是用牛奶、奶油和各种糖果做成的。
10. Bruttiboni is a type of almond-flavoured biscuit made in Prato.	意大利杏仁酥饼干是一种普拉托产的杏仁饼干。
11. I am so excited to share my Tiramisu with you.	能够与你一起分享提拉米苏，令我感到很兴奋。

舌尖聊美食

Conversation 1

Yummy! This is the most flavorful ice cream I had. I love this spumoni with chocolate, vanilla and pistachio.

太美味了！这是我吃过最美味的冰淇淋。我喜欢巧克力味、香草味，还有开心果的意式千层冰淇淋。

Haha! It also meets my taste. But I don't think it is much sweeter than ordinary ice cream or that it can be much too sugary.

哈哈，这种冰淇淋也很合我的胃口。但是我觉得它并不比普通冰淇淋甜多少，否则含糖量就太多了。

Well, that's what an ice cream supposed to be. They must be sweet, otherwise it's weird. Personally, the sugar it contains is acceptable.

冰淇淋本来就应该是甜的啊。不甜就奇怪了。个人觉得，这个冰淇淋含糖量还可以接受的。

Well, in a word, they will make you fat!

总而言之，那会让你长胖的！

Never mind. Look at the juicy fruit layer. Humm, yummy!

没事儿。你看这水果层。真好吃！

开胃词组

much too 太……，非常……

be supposed to 应该，理应

in a word 总而言之

never mind 没关系，不用担心

鲜美单词

flavorful adj. 有味道的，可口的

chocolate n. 巧克力，巧克力糖

pistachio n. 开心果

sugary adj. 甜的，含糖量高的

weird adj. 奇怪的，不寻常的

acceptable adj. 可接受的，令人满意的

juicy adj. 多汁的

layer n. 层，层次

yummy adj. 很好吃的

Conversation 2

Italian desserts typically are rich in flavor. Do you agree with it?	意大利甜品真美味！你同意吗？
Yes, I agree. I have been to Italy, it is hard to say no to all the delicious temptations. It seems that every coffee bar or pasticceria has an endless display of cookies, chocolates or some other enticement.	是的，我同意。我去过意大利。真是难以拒绝那美味诱惑。无论是咖啡店还是糕点店都充满了饼干、巧克力或其他甜品的诱惑。
That's true enough. There are many types of Italian desserts to choose, several sweet treats stand out. Cookies, biscotti, panforte, Milano's panetton, cannoli, and baci are all delicious Italian pastries, desserts and cakes.	没错。有许多意大利甜品可供选择，有几种最为著名，饼干、小圆面包、米兰的潘妮朵尼甜面包、奶油煎饼卷，还有芭喜等等，都是一些意大利面点、甜点和蛋糕。
What's your favorite Italian dessert, anyway?	你最喜欢的意大利甜品是哪种？
Cannoli is definitely my favorite kind. The cannoli is the most famous one of the Sicilian desserts and can be found in every Italian pasticceria. This dessert is made by filling a hollow pastry shell with fresh ricotta.	奶油煎饼卷是我最喜欢的。奶油煎饼卷是西西里最著名的甜品，在意大利各糕点店都有出售。这种甜品是用乳清干酪填到面包片里。
I hear that this dessert is very popular all over the world.	我听说这种甜品在全世界都受欢迎。
At one time, cannoli was a gift given amongst friends during Carnevale, but now has since gained worldwide recognition and numerous delicious variations.	以前，奶油煎饼卷在卡尔内瓦莱是朋友互赠的礼品，但是现在受到全世界认可，品种也在增多。
All right. We're both big fans of Italian desserts.	是的。我们都是意大利甜品的狂热爱好者。

开胃词组

be rich in 含有丰富的，富含

agree with 同意

coffee bar 咖啡馆

stand out 突出，引人注目

all over the world 全世界

at one time 从前

鲜美单词

dessert *n.* 甜点

typically *adv.* 典型地，通常地

temptation *n.* 诱惑，引诱

pasticceria *n.* 糕饼

endless *adj.* 无尽的，无边的

display *n.* 展览，陈列

enticement *n.* 诱惑，怂恿

treat *n.* 招待，款待

panforte *n.* 节日果包

cannoli *n.* 奶油煎饼卷

pastry *n.* 糕点

hollow *adj.* 空的，空洞的

There are so many famous Italian desserts that you can prepare for your parties as an after dinner delight. The most simple of the famous Italian desserts to make yourself is the Tiramisu.Tiramisu is a popular Italian dessert that is so sweet and creamy with a hit of chocolate or coffee making it fabulous! Italian desserts make great holiday dessert recipes for the whole family!

有许多著名的意大利甜品可作为派对餐后的亮点。这些著名的意式甜品当中，最简单、可自己制作的，当属提拉米苏。提拉米苏是受人欢迎的意式甜品，甜美的奶香巧克力或咖啡味，绝妙无比！意大利甜品可谓家庭聚会的甜点精品！

读书笔记

Part 6

French Cuisine
法国情调

Amuse-Bouche 开胃菜

舌尖美食词汇

salad Nicoise 尼古拉斯沙拉	**oeuf cocotte** 法式小盅蛋
pate 馅饼	**verrines** 杯装甜品小蛋糕
gougeres 法式乳酪泡芙	**crab mayonnaise** 蟹肉蛋黄酱
hors d'oeuvres 开胃小菜	**soupe a l'onion** 洋葱汤
foie grat 肥鹅肝	**croissant** 牛角面包

舌尖美食句

1. I recommend this salad Nicoise, a native and famous dish.
   我推荐这个尼古拉斯沙拉，这是当地著名的一道菜。

2. You eat oeuf cocotte by dipping the bread into the cream mix.
   吃法式小盅蛋的时候，要把面包蘸到奶油酱料里。

3. Pate or rillettes are served by spreading on bread.
   吃馅饼和猪肉酱的时候，要把酱料涂在面包层上。

4. In recent years, verrines have become very popular in France for appetizers.
   近几年，杯装甜品小蛋糕作为开胃菜在法国很流行。

5.	Gougères is a tasty treat for dinner party.	法式乳酪泡芙是晚宴聚餐的美味极品。
6.	What would you like? Mushroom feuillete? Crab mayonnaise? Or tart au pistou?	你想要点什么？蘑菇千层酥？蟹肉蛋黄酱？还是香蒜蛋挞？
7.	Hors d'oeuvres are typical French, containing some sea food like mussels and shrimps.	开胃小菜最具法国特色，主要有些海鲜，包括牡蛎和虾。
8.	Soupe a l'onion is my favorite French appetizer.	洋葱汤是我最喜欢的法国开胃菜。
9.	You can't miss foie grat, popular and well-known delicacy in French cuisine.	肥鹅肝是最受欢迎的著名的法国佳肴。
10.	You can make your own croissants at home.	你可以在家自己做牛角面包。
11.	French fish soup takes some beating, usually absolutely full of flavour.	法国鱼汤真让人叫绝，美味十足！

舌尖聊美食

Conversation 1

Wow! How fancy it is, isn't it?	哇！真漂亮，对吗？
Yeah. This French restaurant is full of romance. OK. Let's go with the appetizer.	是啊，这个法国餐厅真富有浪漫气息！好吧，我们该点开胃菜啦。
The French appetizer must be very special.	法国开胃菜肯定很特别。

The first courses are an important part of French cuisine. The common French appetizer includes oeuf cocotte, salade jambon melon, soupe a l'onion, vegetables vinaigrette and so on.

第一道菜是法国菜里相当重要的一环。常见的法国开胃菜有法式小盅蛋、火腿甜瓜沙拉、洋葱汤、什锦蔬菜等等。

How is salade jambon melon?

火腿甜瓜沙拉怎么样?

Well, it is frequently served in the summer in France. Slices of melon mixed with very thinly sliced ham. Delicious!

嗯，这道菜经常在法国的夏天提供。甜瓜片与火腿混搭，很美味!

Is vegetables vinaigrette a kind of vegetable dish?

什锦蔬菜是一道素菜吗?

Bingo! This is one of many types of vegetables, grated or cubed, and tossed in a vinaigrette dressing. The French love grated carrots and beets. Really tasty!

对! 它里面有各种蔬菜，切碎的或方块的，由色拉调味汁杂拌而成的。法国人喜爱碎萝卜和甜菜，很好吃的!

I love vegetables. Can we have a try today?

我喜欢吃蔬菜。我们今天可以试一试吗?

Sure.

当然。

开胃词组

full of 充满

first course 第一道菜

a part of 一部分

have a try 试一试

鲜美单词

romance *n.* 浪漫，温馨

appetizer *n.* 开胃品，开胃小吃

melon n. 甜瓜

frequently adv. 频繁地，屡次地

slice v. 切成薄片

grate v. 磨碎，压碎

cube v. 切成方块

toss v. 轻轻搅拌

beet n. 甜菜

Conversation 2

I'm crazy about French cuisine. It's famous for the rich tastes and elegant style.

我非常喜欢法国菜。法国菜美味又典雅。

I agree with that. And I love French appetizer, such as breads and dips, brochettes and breads.

我同意这一点。我也喜欢法国开胃菜，比如蘸酱面包、小烤串和面点。

Breads and dips are my favorite French appetizer, especially the canape.

蘸酱面包是我最喜欢的开胃菜，尤其是开胃饼。

Is it a special kind of breads and dips?

这是一种特别的蘸酱面包吗？

Well, it's a very common food in France. As a spread, it will normally be on top of a small piece of bread. A triangle of toasted bread with a spread is called a toast, and a more elaborate presentation, featuring several layers, is called a canape.

它是法国常见的一道菜。作为涂酱的一种，通常在小片面包上涂酱料。三角形的吐司面包涂上酱汁就是吐司面包片，如果更复杂一点，涂几层不同的东西的话，就是开胃饼了。

OK. I got it. For French appetizer, I love verrines, quiches, tarts and savory cheese puffs.

哦，我知道了。说到法国开胃菜，我喜欢杯装甜品小蛋糕、乳蛋饼、果馅饼和美味的奶酪泡芙饼。

They do have a lot of appetizers to choose. 他们的确有很多种开胃菜供选择。

That's right. 没错。

开胃词组

be crazy about 对……疯狂，疯狂痴迷

be famous for 因……著名

agree with 同意

such as 例如

鲜美单词

cuisine n. 烹饪，菜肴

taste n. 味道

elegant adj. 优美的，雅致的

favorite adj. 喜爱的

special adj. 特殊的，专门的

common adj. 普遍的，常见的

triangle n. 三角形

spread n. 涂抹在面包上的酱

quiche n. 乳蛋饼

tart n. 果馅饼

savory adj. 好吃的

For the French, meals are always very organized. All meals are important for them and they follow a particular pattern while planning and eating their food. You will be surprised to know, that appetizers are eaten by the French, even when they are grabbing a quick lunch at home or work. It is immaterial, whether the person is eating alone or in a group. The French call them entree. Entree, translated into English, stands for entrance. No doubt that they are indeed an entrance to the meals. However, for most Americans, an entree stands for the main course. Most French find it difficult to adjust to this concept, when they are visiting the United States of America.

法国人用餐非常讲究，每一餐对于他们来说都非常重要，他们用餐遵循特定的流程。你会惊奇地发现，法国人无论家庭用餐还是工作用餐，都非常注重开胃菜。无论一个人用餐还是多人一起用餐，他们都点开胃菜。法国人称之为头盘菜。头盘菜，“entree”就是“入口，入场”的意思。难怪“头盘菜”确实是用餐的“入场”菜。然而，大多数美国人都称“entree”为主菜。法国人在美国很难把这两个概念转换过来。

An appetizing drink is served at the beginning of the meal which is known as aperitif. It is often an alcoholic drink which aims to increase the appetite of the person. The drink is accompanied by some light snacks. Normally, nuts or olives are served.

在用餐开始的时候，法国人常常会饮用开胃酒，正是所谓的“餐前酒”。这是一种含酒精的酒，旨在增进用餐者的胃口。饮用开胃酒的时候，通常吃些小点心，比如坚果仁或橄榄等等。

UNIT 02

Bread
面包主食

舌尖美食词汇

French bread 法式面包	**baguette** 法式长棍面包
French toast 法式吐司	**croissant** 羊角面包
brioche 黄油鸡蛋圆面包	**ficelle** 细条面包
country bread 法式乡村面包	**faluche** 贝雷面包
bread pudding 面包布丁	**French bread rolls** 法式面包卷

舌尖美食句

1. We all love freshly baked French bread, with its soft and fluffy texture and crisp golden crust.
   我们都喜欢新鲜出炉的法式面包，里面柔软蓬松，外面金黄酥脆。
2. Baguette is made from wheat flour, water, yeast and common salt.
   法式长棍面包是由面粉、水、酵母粉和食盐做的。
3. Would you like some French toast for the breakfast?
   早餐你想吃法式吐司吗？
4. She loves the crusty and chewy croissant.
   她喜欢脆皮又耐嚼的羊角面包。

5.	Brioche is rich in sweet flavor and flaky texture.	黄油鸡蛋圆面包富有甜味且脆嫩。
6.	You must eat the ficelle as soon as possible, because they are so thin that the inside dries out rather fast once they have been baked.	细条面包要尽快吃，因为它们太细小，烤出来之后，里面很快变干变硬。
7.	I use an overnight starter to give my country bread extra flavor.	我用隔夜酵母给法式乡村面包增加不同的口味。
8.	Faluche is a pale white bread that is soft and fairly dense.	贝雷面包是一种又软又厚实的灰白面包。
9.	Pain brie is a traditional Normandy bread.	布里面包是一种诺曼底的传统面食。
10.	We have bread pudding in various flavors.	我们有各种口味的面包布丁。

舌尖聊美食

Conversation 1

What are you eating for the breakfast?	你早餐在吃什么呢？
A French stick, crusty and chewy, yum!	一根“法式长棍棒”，脆皮又耐嚼，好吃！
Sorry! I didn’t get it. What’s “French stick”?	不好意思！我不明白。什么是“法式长棍棒”？
It’s a kind of French bread. It is known as a "baguette", which literally means "a stick".	它是一种法式面包，名叫法式长棍面包，字面意思就是“长棍”。

 I don't know much about French bread. Can you tell me more about it?	我不太了解法式面包。你能给我介绍一下吗?
 Sure. French bread is a very common staple food for French people. In addition to baguettes,France has a wonderful range of delicious white breads to offer. French white bread comes in several other shapes and sizes, for example, couronne, flute, batard and ficelle.	当然。法式面包是法国人常见的主食。除了长棍面包，法国还有很多美味的白面包。法式白面包，根据不同的形状和大小，有法式花边面包、长笛面包，还有巴塔面包和细条面包。
 White bread?	白面包?
 Well, white bread is bread made from white wheat flour. Apart from these basic types of bread,France's bakeries also sell a whole range of other types of bread made in whole wheat flour or some rye flour, including wholemeal breads, rye bread, sourdough bread and a sweet bread called brioche.	嗯，白面包就是白面粉做的面包。除了这些常见的面包，法国面包师还制作许多其他种类的面包，用全麦粉或黑麦粉做的，包括粗面粉面包、黑麦面包、酵母面包和黄油鸡蛋圆面包。
 It sounds amazing. They must be delicious.	听起来很神奇，这些面包一定很好吃。

开胃词组

get it 懂得，了解

be known as 被认为，号称

staple food 主食

in addition to 除……之外

be made from 由……所做成

apart from 除此之外，除去

鲜美单词

breakfast *n.* 早餐，早饭

crusty *adj.* 硬皮的，脆皮的

chewy *adj.* 耐嚼的

baguette *n.* 法国长棍面包

staple *n.* 主食

range *n.* 范围，类别

couronne *n.* 花边外缘饰圈

flour *n.* 面粉

wholemeal *n.* 全麦面粉

rye *n.* 黑麦

sourdough *n.* 酵母

Conversation 2

Can I help you?	你想买什么？
I'd like two loaves of French country bread, please.	我想买两条法式乡村面包。
OK. That's 20 dollars in total.	好的。一共20美元。
Hmm, they're quite expensive. Any discount?	有点贵啊。有打折吗？
Sorry, this is the bottom price. Why don't you consider Combo Three? It includes two country bread, one French toast and a submarine sandwich.	对不起，这是最低价。你要不考虑一下三号套餐？里面有两个乡村面包、一个法式吐司，还有潜艇三明治。
What's the price?	多少钱？
It only costs you 20 dollars.	只需要20美元。

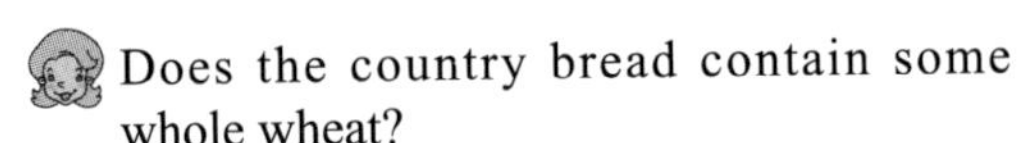

Does the country bread contain some whole wheat?

这乡村面包是全麦的吗？

Yes, it has some whole wheat flour or some rye flour inside, so that you can keep it longer.

是的，它内含全麦粉和黑麦粉，这样就可以保存长久一些。

Well, I will take this combo.

好，我买这个套餐。

开胃词组

in total 一共，总共

bottom price 最低价格

whole wheat 全麦

so that 以便；结果，以致

鲜美单词

total n. 总计，总数

expensive adj. 贵的，昂贵的

discount n. 折扣，减价

bottom adj. 底部的，末端的

combo n. 套餐

toast n. 吐司，烤面包片

submarine n. 潜艇

cost v. 需付费，价钱为

flour n. 面粉

Brioche is a French bread, characterized by a rich, sweet flavor and flaky texture that almost makes it feel like a pastry, rather than a bread. It is often served as a breakfast food and can be sweet or more savory, depending on how the recipe is manipulated. The sweeter versions are sometimes confused with cake, but they really are bread, since they are leavened with yeast and kneaded. Many French bakeries sell this bread, and it can also be made at home.

黄油鸡蛋圆面包是一种法式面包，具有香甜可口、外酥里嫩的特点，吃起来像糕点，而不像面包。它主要在早餐时候食用，制作时可根据食谱来决定它的甜度。最甜的黄油鸡蛋圆面包可比肩蛋糕，但是它们的确还是一种面包，因为制作它们的时候放了酵母，并揉捏面团。许多法国面包店都有销售这种面包，但它也可以在家制作。

The term “brioche” dates back to at least the 1400s, and the bread itself may be even older. The word is a name for a style of bread and dough that can be made in a number of shapes. A classic shape has a fluted bottom and an upper protruding knob, and is made in a special brioche pan. The bread can also be made like a regular loaf, or it can be braided or molded into a ring. It often takes the form of an individual bun, served warm.

黄油鸡蛋圆面包的来历可追溯到15世纪，这种面包本身也许更古老。它的名字起源于面包或面团的许多形状。最经典的形状就是底部凹陷，顶部凸出的那种，并在特殊的烤盘里制成。这种面包还可以做成普通的形状，也可以做成环状的，小圆面包的形状更常见，最好热的时候食用。

Sweet varieties can be filled with things like fresh or candied fruit or chocolate. The dough may be given some extra sweetness to complement the filling, making the dish particularly decadent. Savory versions are filled with vegetables or meats, depending on the taste of the cook. In all cases, it has a light, flaky crust with a high gloss, caused by brushing the dough with egg before baking.

黄油鸡蛋圆面包在制作的时候可以放一些甜料，比如甜果或巧克力。面团也额外放些糖作为补充，让味道更厚重些。根据各种不同口味，需要放蔬菜、肉类等等。所有黄油鸡蛋圆面包在制作的时候，都在面团外层涂上了鸡蛋，让它们烤出来带有一层薄薄的脆皮光泽。

读书笔记

UNIT 03

A Nutritious Breakfast 营养早餐

舌尖美食词汇

ice cream 冰淇淋	**cacao powder** 可可粉
fruit juice 果汁	**hot chocolate** 热巧克力
bread roll 面包卷	**cereal** 麦片
yogurt 酸奶	**apricot jam** 杏仁酱
cookies 饼干	**café au lait** 牛奶咖啡
tartine 奶油果酱面包	

舌尖美食句

1. The French love to eat baguette for their breakfast.	法国人早餐喜欢吃法式长棍面包。
2. Have you tried to bake the croissants with chocolate?	你尝试过烤羊角面包的时候放巧克力吗？
3. Some fruit juice with Fresh bread would be perfect for me.	对我来说，果汁和新鲜面包是完美的搭配。
4. This restaurant serves hot chocolate every morning.	这个餐厅每天早上都提供热巧克力。

5. These are some of the best bread rolls I've ever had.	这些是我吃过的最好的面包卷。
6. Cereal is very healthy food for breakfast.	麦片是健康的早餐食物。
7. Sometime I also drink some yogurt in the morning.	有时候早上我也喝些酸奶。
8. I prefer spreading on the bread some honey, strawberry or apricot jam.	我喜欢在面包上涂蜂蜜、草莓或杏仁酱。
9. These cookies taste like a treat.	这些饼干很好吃。
10. Usually, café au lait is for the morning.	牛奶咖啡通常在早上喝。
11. I would like a small yoghurt and one or two tartines with jam for my breakfast.	作为早餐，我喜欢小杯酸奶，还有涂酱的一两个奶油果酱面包。

舌尖聊美食

Conversation 1

Good morning, Amy! Why do you walk in a rush?	早上好，艾米！你为什么走得那么匆忙？
Good morning. I'm going to get in the line.	早上好。我要去排队。
What?	怎么回事？
Well, I mean I go to buy my breakfast. In France, people usually eat baguette for breakfast. They will be sold out if you're late.	嗯，我是说我要去排队买早餐。在法国，人们早餐喜欢吃长棍面包。去晚了，就卖光了。
Oh, yes. But I prefer croissant to baguette for my breakfast.	哦，是的。但与长棍面包相比，我更喜欢羊角面包。

What about the drink?	饮料喝什么啊?
I'm apt to some black coffee or hot chocolate.	我喜欢喝黑咖啡或热巧克力。
If I can't get baguette, I like something nutritious for my breakfast, cereal and yogurt.	我要是买不到长棍面包，我喜欢吃些有营养的东西，麦片和酸奶什么的。
 I agree with you. The French breakfast is usually light and healthy.	我赞同。法国早餐一般都清淡，很健康。

开胃词组

in a rush 匆匆忙忙

sell out 售完，卖光

prefer… to 更喜欢……

be apt to 往往，倾向于

agree with 同意，赞同

鲜美单词

rush *n.* 仓促，匆忙

line *n.* 队，队伍

breakfast *n.* 早餐，早饭

baguette *n.* 法式长棍面包

apt *adj.* 有……倾向的，易于……的

coffee *n.* 咖啡

chocolate *n.* 巧克力

nutritious *adj.* 有营养的，营养丰富的

cereal *n.* 谷类食物，麦片

healthy *adj.* 健康的

Conversation 2

Hey. Charlie. Since you were born in Paris, may I ask what a traditional French breakfast is?

嘿，查理！你是法国人，所以问一下传统的法国早餐都吃什么呀？

Well, French breakfast is kind of light. For breakfast we eat various breads and drink coffee.

法国早餐有点清淡，我们通常吃各种面包，再喝点咖啡。

What a surprise! I thought you eat the same thing as the British, like sausage, ham, eggs and bacon.

真奇怪！我原以为你们也像英国人那样早餐吃香肠、火腿、鸡蛋还有培根。

Actually, a traditional French breakfast is not as heavy as other European breakfast meals.

实际上，传统法国早餐不像其他欧洲国家的那样。

What's the detail?

具体来说呢？

On a regular day, people eat slices of French bread served with jam or butter. The breads include baguette, bread rolls with chocolate filling and freshly-baked croissants. Croissants are often eaten plain, with jam, or dipped in coffee bowls. However, if you want to be truly French, never eat a croissant with butter, because French people never do that.

每天早上，人们蘸着酱或黄油，吃几片法国面包。包括长棍面包、带巧克力的面包卷或新鲜出炉的羊角面包。羊角面包要蘸酱吃，或蘸着咖啡吃。而如果要像法国人那样吃羊角面包，千万别涂上黄油，因为法国人从来不这么吃。

What about the drink?

那饮料呢？

To go with bread, French people usually drink coffee in the morning, and children are allowed to drink hot chocolate or fruit juice instead.

早餐吃面包，法国人时常喝些咖啡，小孩们则喝热巧克力或果汁。

That's it?

就这些？

Well, sometimes we eat some fruits or yogurt, too.

有时候我们也吃水果和酸奶。

That sounds very healthy to start the day.

听起来真是以非常健康的早餐开始新的一天。

开胃词组

be born in 出生于

kind of 稍微，有点儿

the same as 与……相同

as heavy as 与……一样重

鲜美单词

traditional adj. 传统的，习俗的

various adj. 各种各样的

sausage n. 香肠

bacon n. 培根，熏猪肉

European adj. 欧洲的，欧洲人的

slice n. 片，薄片

plain adj. 朴素的，简朴的

dip v. 蘸，浸

bowl n. 碗，盆

allow v. 允许，容许

What do French people usually take for breakfast?

法国人早餐吃什么?

In France, adults are inclined to take a bowl, a mug or a cup of very strong coffee. This coffee often goes with French bread called "baguette". As for the drink, some adults love to dilute their coffee with hot milk and we call this mixture a "café au lait". Other people enjoy better something sweeter in the morning with their beverage and would prefer spreading on their bread some honey, strawberry or apricot jam. Brioche, pain au lait (slightly sweet light bread) and biscuits are other alternative that can also be part of the French breakfast.

在法国，成年人喜欢喝一马克杯或一小杯浓咖啡。通常喝咖啡时，吃些法式长面包。在饮料方面，他们喜欢在咖啡里加些热牛奶，这种饮品叫"热牛奶咖啡"。其他人，早餐也许喜欢吃些甜品，喝些饮料，还有涂着蜂蜜、草莓酱或杏仁酱的面包片。法式早餐，也有奶油蛋卷、牛奶面包和饼干等可供选择。

Children do not drink coffee and rather favour their hot chocolate made by pouring some cacao powder into some hot milk. Many French people and especially children also like to soak or dunk their slices of bread into their bowl of hot chocolate. Other children would just have some cereals with milk only. Tea which is often part of a British breakfast is a beverage that is rarely consumed in a French breakfast although in some French families it can be taken especially by children.

孩子们不喝咖啡，他们喜欢加点可可粉和牛奶的热巧克力。许多法国人，尤其孩子们喜欢蘸着热巧克力吃面包片。有的孩子只吃点加奶的麦片。喝茶是英式早餐的传统，而法国人早餐很少喝茶，尽管有的法国家庭，尤其孩子们也会早上喝茶。

UNIT 04 A Good Lunch 丰盛午餐

舌尖美食词汇

crevetters sauce boursin 乳酪海鲜	**French bread** 法式面包
blanquette de Veau 白汁烩小牛肉	**saumon grille** 烤三文鱼
foie gras 鹅肝酱	**mussels with roquefort** 羊干酪贻贝
chocolate mousse 巧克力慕斯	**caviar and steak tartare** 鱼子酱鞑靼牛排
Lobster Au Gratin 芝士焗龙虾	**French beef stew** 法式炖牛肉

舌尖美食句

1. Crevetters sauce boursin is not a suitable dish for those who have allergy to shrimp.	乳酪海鲜这道菜不适合对虾过敏的人食用。
2. French bread and a red wine would also be a must for the lunch.	法式面包和红酒是午餐必点的食物和酒饮。
3. Blanquette de Veau is a traditional, simple homemade dish.	白汁烩小牛肉是一道传统的、简单的家常菜。

4. If you don't have enough time to prepare, saumon grille could be a good choice.	如果没有足够的时间烹饪，烤三文鱼或许是一个不错的选择。
5. You will not miss foie gras, the classic one for the appetizer.	你不会错过鹅肝酱的，经典的开胃菜。
6. Mussels with Roquefort becomes a true feast if you serve it with a red wine.	羊干酪贻贝与红酒搭配是一道真正的盛宴。
7. Chocolate mousse is a great dessert for entertaining.	巧克力慕斯是一个非常棒的甜品。
8. How do you cook monkfish?	你怎么做琵琶鱼呢？
9. I'll prepare the caviar and steak tartare for you.	我给你准备鱼子酱鞑靼牛排。
10. Lobster Au Gratin is a gourmet food, a traditional French cuisine.	芝士焗龙虾是一道法国传统美食。
11. French beef stew is a flavorful, full-bodied low oxalate stew.	法式炖牛肉是一道非常美味的、低草酸的、味道醇厚的炖肉菜。

舌尖聊美食

Conversation 1

Excuse me, sir. Can I take your order now?	打扰一下，先生。我能帮您点菜了吗？
Yes. I would like to order some authentic French cuisines. Can you give me some recommendations?	好的。我想点一些正宗的法国菜。你能给点推荐吗？

Sure. For the appetizer, I recommend crevetters sauce boursin, a popular dish for the first course. It contains shrimp sauteed with sun-dried tomatoes, corn and leeks in a garlic, herb cream sauce.	当然。对于开胃菜，我推荐乳酪海鲜，受欢迎的第一道菜。这道菜是用咖喱酱和奶油酱浇炸土豆、海鲜虾、玉米与香葱做成的。
OK. I'll take it. I love seafood. What about the main course?	好，我点这个，我喜欢海鲜。主菜呢?
Well, Blanquette de Veau is our specialty. This is a classic "white stew", ivory veal stew with mushrooms, served over rice. It could be a good option for the main course.	嗯，白汁烩小牛肉是我们的特色菜。这是经典的炖白肉，乳酪炖牛肉和蘑菇，与米饭一起吃。这是不错的主菜之选。
I think it is a bit light. I love something hearty and savory.	我觉得这有点清淡了。我喜欢丰盛美味的。
Sorry, sir. French cuisine is not as heavy as other countries', like American cuisine. But I strongly recommend this saumon grille, herbed grilled salmon. It might be suitable for you.	对不起，先生。法国菜没有其他国家的菜式那样重口味，比如美国菜。但是我强烈推荐这道香草烤三文鱼。这道菜也许适合你。
OK, it sounds delicious. I will take it. And for the dessert, I would like an ice cream.	好，听起来很美味。我就点这个了。至于甜点，我想来个冰淇淋。
Fine. I will come back right away.	好的，我马上就回来。

开胃词组

take order 点菜，写菜单

the first course 第一道菜

the main course 主菜

be suitable for 适合，适于

right away 马上

鲜美单词

order n. 点菜

authentic adj. 真的，真正的

cuisine n. 烹饪；菜肴

recommend v. 推荐

shrimp n. 虾，小虾

leek n. 香葱

seafood n. 海鲜

specialty n. 特色食品

veal n. 牛肉，小牛肉

mushroom n. 蘑菇

hearty adj. 丰盛的

savory adj. 好吃的

Conversation 2

What's a typical French lunch?

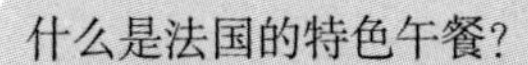
什么是法国的特色午餐？

Generally, we start with salad with potatoes or cucumbers. For the entree, we take beefsteak, chicken and fish. French people also eat a slice of Camembert with fresh baguette, and a fancy little French cake like opéra, tryanon or Paris-Brest. Then, drink one or two glasses of red wine and water with the dessert.

一般来讲，我们开胃菜吃土豆沙拉或黄瓜沙拉。主菜方面，我们吃牛排，鸡肉和鱼肉。还吃卡门贝尔面包和美味的法式蛋糕，比如剧院蛋糕，特里亚农或巴黎车轮饼。然后喝一两杯红酒和水。

I mean what the exact dishes you eat, for example, you may eat basil salmon terrine for the appetizer.

我是说你们吃什么具体的菜，比如开胃菜你们可能吃鲑鱼面包片。

Well, we usually have bisque, foie gras, French onion soup for the appetizer. And for the entree, we have cassoulet, steak tartare or mussels. Typical patisserie, crepe, mousse au chocolate are perfect for the desserts.

哦，作为开胃菜我们通常喝法式浓汤，吃鹅肝酱，法式洋葱汤。主菜方面，什锦砂锅，牛肉鞑靼，贻贝是特色之选。特色糕点，可丽饼，巧克力奶油冻等是点心极品。

Wow, they're all tasty delicacy.

哇，全都是可口的美食呀。

Yes, they are. They are also healthy and refined.

是的，这些美食健康又精致。

开胃词组

red wine 红葡萄酒

for example 例如

鲜美单词

typical *adj.* 典型的

generally *adv.* 一般地，通常

entree *n.* 主菜

Camembert *n.* 卡门贝尔（法国地名）

cassoulet *n.* 砂锅炖肉豆

tartare *n.* 鞑靼牛肉

mussel *n.* 贻贝，蚌类

patisserie *n.* 法式蛋糕

delicacy *n.* 珍馐；佳肴

refined *adj.* 精炼的，精致的

French beef stew is traditional served with steamed or boiled potatoes. It's also great with a Potato Gratin Cooking the beef slowly and gently in plenty of liquid is a great way to deal with the sometimes tougher and more strongly flavored nature of grass-fed beef, especially for people first making the switch to truly grass-fed beef. Stewing breaks down the toughness of any meat, and the acid in the wine also stands up against the stronger, beefier flavor some people need to ease into when they try grass-fed meat.

法式炖牛肉是一道传统法国菜，与土豆一同食用，与脆皮土豆搭配也很棒。炖牛肉的时候，需要在汤里慢慢煨炖，这是一种让更柔韧的食草牛肉煮出纯美味道，尤其对于那些喜欢食草牛肉正宗味道的人们。煨炖之后，肉质鲜嫩，红酒中的酒石酸也能混合浓厚的牛肉味。但有的人吃牛肉的时候，需要减轻这种牛肉味。

读书笔记

A Fancy Dinner
尚品晚餐

舌尖美食词汇

Macaron 马卡龙杏仁甜饼	**steak Diane** 戴安牛排
cherry clafouti 樱桃沙司饼	**chateaubriand** 烤里脊牛肉
Custard 蛋奶沙司	**ratatouille crepes with goat cheese** 羊乳酪蔬菜饼
roasted rack of lamb 烤羊排	**oyster fritters** 面拖牡蛎
chocolate mousse 巧克力慕斯	**pan-seared sea scallops** 香煎鲜贝

舌尖美食句

1. Would you like to have some macarons for the dessert?
你来点马卡龙杏仁甜饼做点心吗？

2. Is this steak Diane special?
戴安牛排很有特色吗？

3. This cherry clafouti is a super sweet mix of custard, cherries, and a light, flaky crust.
樱桃沙司饼是由蛋奶沙司、樱桃和薄面饼做成的，非常甜美。

4. Chateaubriand is a classic French dish.
烤里脊牛肉是一道经典法国菜。

5. How's your custard, sweet or salty?	蛋奶沙司吃起来怎么样，甜的还是咸的？
6. Ratatouille crepes with goat cheese are a bit time-consuming dish but it offers a great flavor.	羊乳酪蔬菜饼制作过程很费时，但吃起来很有风味。
7. Would you like to try this roasted rack of lamb?	你想尝尝这份烤羊排吗？
8. These oyster fritters are made with a thick batter and lots of chopped oysters.	这些面拖牡蛎是用大块面团和牡蛎碎肉做成的。
9. This rich and decadent chocolate mousse is easy to make.	这份美味浓稠的巧克力慕斯很容易制作。
10. I like to order some pan-seared sea scallops.	我想点香煎鲜贝。
11. Sweetbreads refer to various glands used as food.	杂碎是指各种作为食物的动物胰脏。

舌尖聊美食

Conversation 1

Hey, Cindy! We will have a fancy dinner party on Saturday night. Would you like to come?	嘿，辛迪！我们周六晚上举行盛大的晚宴派对，你来吗？
Great! Should I bring a homemade dish?	好啊！需要带自己做的菜吗？
This is a homemade food party again. What will you prepare for that?	这又是一个家庭家常菜派对。你准备带什么呢？

I think roasted rack of lamb will be fantastic treat for everyone.	我想烤羊排大家都会喜欢吧。
It must be an authentic French style delicacy.	那肯定是一道正宗的法国风味佳肴了。
Yes. Rack of lamb is usually roasted, sometimes first coated with an herbed breadcrumb persillade. The persillade crust is very versatile and can be used for eggplants, tomatoes or even roast cod.	没错。烤羊排是一种烤肉，有时外皮有一层面包屑脆皮。这层脆皮有很多种，可以用茄子、番茄甚至烤鳕鱼来做。
It sounds delicious. I'm looking forward to it.	听起来十分美味哦，我很期待。

开胃词组

homemade dish 家庭自制菜，家乡小菜

prepare for 为……而准备

look forward to 期待，期望

鲜美单词

fancy adj. 精致的，精美的

homemade adj. 自制的，家庭自制的

prepare v. 准备，预备

rack n. 支架；（羊、猪等带前肋的）颈脊肉

authentic adj. 真的，真正的

delicacy n. 美味佳肴

breadcrumb n. 面包屑

crust n. 面包皮

versatile adj. 多功能的，多种用途的

cod n. 鳕鱼

Conversation 2

Hi, what do you want to have, Sir.?	嗨，你想点什么，先生？
Could you recommend some typical French dishes for dinner?	你可以推荐一些特色法国菜给我吗？
No problem. Well, do you have any allergy to seafood?	没问题！嗯，你对海鲜不过敏吧？
No. Actually, I love to eat seafood.	没有。实际上，我喜欢吃海鲜。
How about oyster fritters for the appetizer, roasted lemon rosemary chicken for the main dish, chocolate mousse and cinnamon orange crepes for the dessert?	开胃菜来个面拖牡蛎，主菜来个香烤柠檬鸡，巧克力慕斯和肉桂橙可丽饼做点心，怎么样？
What is in oyster fritters?	面拖牡蛎里面有什么？
Hmm, it contains oyster meat and wine batter. What a treat it is! Naturally salty oyster meat is coated with a savory sparkling wine batter and deep-fried until it turns hot and crispy.	面拖牡蛎里面有牡蛎肉和酒面团。很美味的！它主要由酒面团覆盖在牡蛎肉上面，油炸变脆，然后上桌。
Is this roasted lemon rosemary chicken special?	这个香烤柠檬鸡有什么特别的吗？
The chicken meat is roasted with onions, carrots, celery, garlic juice on top and rub the lemon zest, parsley, rosemary, and thyme under the skin.	这道菜鸡肉是用洋葱、胡萝卜、香芹、蒜末做成的酱汁淋在鸡肉上，在鸡肉皮擦上柠檬皮、欧芹、迷迭香和百里香，然后烘烤烹制的。
Good, I can't wait for this yummy treat. Oh, I would like to order a bottle of fine red wine to go with my dinner.	好，我等不及这丰盛的晚餐啦。哦，我要点一瓶上等红葡萄酒配晚餐。
Sure. I will be right back to you.	好的。我马上为您端上来。

开胃词组

no problem 没问题

have allergy to 对……过敏

main dish 主菜

wait for 等待

鲜美单词

allergy *n.* 过敏反应，过敏症状

oyster *n.* 牡蛎

fritter *n.* 油炸馅饼

rosemary *n.* 迷迭香

mousse *n.* 奶油冻，慕斯

cinnamon *n.* 肉桂

crepe *n.* 薄煎饼

coat *v.* 给……涂上，裹上

sparkling *adj.* 起泡沫的

crispy *adj.* 脆的，酥脆的

rub *v.* 擦，涂抹

舌尖美食文化

Coq au vin is a French dish of chicken braised with wine, lardons, mushrooms and optionally garlic. While the wine used is typically Burgundy, many regions of France have variants of coq au vin using the local wine, such as coq au vin jaune (Jura), coq au Riesling (Alsace), coq au pourpre or coq au violet (Beaujolais nouveau), coq au

Champagne, etc.

红酒焖鸡是一道法国菜，鸡肉与红酒、腊肉、蘑菇，还有大蒜等一起烹制。红酒一般用法国勃艮第红酒。法国各地区的红酒焖鸡也用当地红酒来烹制，比如法国汝拉地区的黄葡萄酒焖鸡、法国阿尔萨斯地区的雷司令酒焖鸡、法国博若莱新葡萄酒焖鸡、香槟酒焖鸡等等。

Coq au vin was created as a delicious way to tenderize a tough, old bird in poor households. Chicken stewed in wine is a wonderful, hearty meal that needs no more than a baguette and good wine to be complete.

红酒焖鸡以特殊的方式炖出鲜嫩的柴鸡肉，味道鲜美、浓厚，食用的时候，无须长棍面包或好酒搭配。

This recipe is prepared country-style, because the stewing vegetables are not removed upon serving, and the bacon, pearl onions, and mushrooms are partially cooked right along with the chicken.

这道菜以乡村传统烹饪法制作，蔬菜铺在上层，培根、小洋葱和蘑菇与鸡肉在下层炖制而成。

读书笔记

Delicious Soups 鲜美羹汤

舌尖美食词汇

French onion soup 法式洋葱汤	**French garlic soup** 法国大蒜汤
daube de boeuf 汤炖牛肉	**garbure** 腌肉菜汤
chestnut soup 栗子汤	**soupe au pistou** 蔬菜蒜泥浓汤
shrimp bisque 鲜虾浓汤	**chicken sausage gumbo** 鸡肉腊肠秋葵汤
lobster bisque 龙虾浓汤	**potato and leek soup** 土豆香葱汤
chilled beet soup 甜菜冷汤	

舌尖美食句

1. French onion soup is usually a plate that is served as a starter.
法式洋葱汤通常作为一道开胃菜。

2. French garlic soup comes from the Languedoc region of France.
法国大蒜汤来自于法国朗格多克地区。

3. Daube de boeuf is a lovely stew.
汤炖牛肉味道不错。

4.	Garbure is a thick meat, bean and vegetable soup.	腌肉菜汤是一种肉、豆、蔬菜杂烩浓汤。
5.	Chestnut soup is rich, creamy and full of winter spices, perfect for a cold day.	栗子汤美味圆润，有许多冬季调味料，很适合冷季节食用。
6.	Soupe au Pistou is basically a seasonal vegetable soup with pistou sauce.	蔬菜蒜泥浓汤是一道由蒜泥酱制成的季节性蔬菜汤。
7.	Shrimp Bisque is a very quick and simple dish, full of flavour!	鲜虾浓汤是一道简单快捷的汤品，很有风味！
8.	Chicken sausage gumbo is cooked in the slow cooker with okra, smoked sausage, green pepper, celery and other seasonings.	鸡肉腊肠秋葵汤，用秋葵、熏腊肠、绿椒、芹菜和其他配料在锅里慢慢烹制而成。
9.	Lobster bisque soup is the perfect appetizer for any dinner.	龙虾浓汤非常适合作为晚餐开胃菜。
10.	She loves potato and leek soup.	她喜欢土豆香葱汤。
11.	We are going to prepare chilled beet soup for dinner.	我们打算准备甜菜冷汤作晚餐。

舌尖聊美食

Conversation 1

Welcome to my house, Amy.	欢迎到我家来，艾米！
Thank you very much for inviting me.	谢谢您的邀请！
I recommend to you a kind of authentic French soup, French onion soup. Please take a sip of it.	我向你推荐一道正宗法国汤……法式洋葱汤。请你先尝一口。

OK. Mmm, yummy! What's with this soup?

好的。嗯，味道美极了！这汤里有什么？

Well, French onion soup is a type of soup usually based on meat stock, and often served gratineed with croutons and cheese on top.

嗯，法式洋葱汤是一种高汤，上面还有面包片或奶酪等。

How do you cook it?

您是怎么做这道菜的呢？

It's quite simple. First, saute the onions in the olive oil in a large saucepan, slowly cook and caramelise the onions with sugar and garlic and saute. Then add the stock, vermouth or wine, bay leaf and thyme. Simmer them until the flavors are well blended, about 30 minutes. Finally, cover with the toast and sprinkle with cheese. Put into the broiler for 10 minutes until the cheese bubbles and is slightly browned. Serve immediately.

这道菜烹饪很简单的。首先，用橄榄油在炖锅里慢慢煎洋葱，加点糖、大蒜和酱油，让洋葱变得有点焦红。然后加入肉汤、苦艾酒或其他酒、月桂叶和百里香。煨炖30分钟，让它们的味道融合一起。最后，在汤的上面放一块吐司面包片或一块奶酪。放入烤箱里烤10分钟，让奶酪冒泡或者变棕黄色，就可以吃了。

What's the key to the fantastic taste?

做出这种美妙味道的关键是什么？

The key to a great French onion soup is starting with good stock. Another important element is the proper caramelization of the onions. The browning, or caramelizing, of the onions brings out the sweetness in them.

上等法式洋葱汤的关键是要有好的肉汤，另一个关键因素是洋葱焦化得好。洋葱变焦的过程中也焖出了甜甜的味道。

OK, I see. Thank you.

好的，明白了。谢谢您！

It's my pleasure.

这是我的荣幸。

开胃词组

take a sip of 喝一口

base on 基于，建立在……基础之上

the key to 关键是……

bring out 呈现出；激发

鲜美单词

invite *v.* 邀请

sip *v.* 小口地喝，抿

stock *n.* 高汤，原汤

crouton *n.* 油炸面包丁，吐司丁

olive *n.* 橄榄

saucepan *n.* 炖锅，深平底锅

vermouth *n.* 味美思酒，苦艾酒

simmer *v.* 用文火炖，煨

blend *v.* 混合

sprinkle *v.* 撒，洒

caramelize *v.* 使变成焦糖

Conversation 2

Good afternoon. Do you provide room reservation?

下午好，你们提供房间订餐服务吗？

Yes. What would you like?

有，你想点什么？

What kind of soup do you have?

你们有什么汤？

We have oyster soup, lobster bisque, chestnut soup and celeriac parsnip soup.	我们有牡蛎汤、龙虾浓汤、栗子汤、芹菜萝卜汤。
Oh, I see. Lobster bisque, please. And a sandwich.	哦，知道了。那就龙虾浓汤吧，还有来一个三明治。
OK. Do you prefer something else?	好的，还要什么吗？
Well, not really. Could you get everything ready in 10 minutes? I'm sort of hungry.	没了。你们10分钟内能做好吗？我有点饿了。
Sure. What's your room number?	当然。您的房间号码是多少？
Room 1102.	1102房间。
OK. We are coming soon.	好的。我们马上就来。

开胃词组

get ready 准备

sort of 有点，有几分

room number 房间号码

鲜美单词

reservation n. 预约，预订

lobster n. 龙虾

bisque n. 浓汤

chestnut n. 栗子

celeriac n. 根芹菜

parsnip n. 欧洲萝卜，欧洲防风草

prefer v. 更喜欢，偏好

Traditionally, French soups are classified into two groups: clear soups and thick soups. The established French classifications of clear soups are bouillon and consomme. Thick soups are classified depending on the type of thickening agent used: purees are vegetable soups thickened with starch; bisques are made from pureed shellfish thickened with cream; cream soups are thickened with bechamel sauce; and veloutes are thickened with eggs, butter and cream. Other ingredients commonly used to thicken soups and broths include rice, flour, and grain.

根据传统，法国汤品分为两类：清汤和浓汤。法国的清汤主要包括肉汤和清炖肉汤。浓汤依据添加的佐料可分为：用淀粉稠化的蔬菜汤，用奶油稠化的海鲜浓汤，贝沙梅尔沙司酱稠化的奶汤，还有鸡蛋、黄油、奶油稠化的浓汤。其他用来使汤汁变稠的佐料还有米饭、面粉和其他谷物。

Fresh ingredients are the basics to really tasty French soups. As the climate and landscape changes throughout France, regional favourites depend on what is available around the seasons.

新鲜的原材料是法国汤品的基本要素。法国各地气候变化不同，让各地区美食一年四季拥有不同的风味。

读书笔记

Special Desserts 特色点心

舌尖美食词汇

souffle 蛋奶酥	creme caramel 焦糖布丁
rhubarb tart 大黄塔	praline crepe 果仁薄饼
Financier cake 金砖蛋糕	clafouti 法式樱桃布丁
chocolate fondant 巧克力沙翁	pastry dough 油酥面团
chocolate mousse 巧克力慕斯	waffle 华夫饼

舌尖美食句

1. Did you make these souffles by yourself? 这些蛋奶酥是你自己做的吗?
2. Could you pick up some creme caramel for me? 你能给我挑选一些焦糖布丁吗?
3. This rhubarb tart tastes sweet and creamy. 这个大黄塔甜甜的，很嫩滑。
4. Who can resist a tasty praline crepe? 谁能抵挡果仁薄饼的诱惑呢?

5. Financier cake is light and moist, similar to sponge cake.	金砖蛋糕味道清淡，润滑，有点像海绵蛋糕。
6. Would you like to try clafouti?	你想试试法式樱桃布丁吗？
7. Chocolate fondant is a lovely treat.	巧克力沙翁是很可口的甜品。
8. I love to eat pastry dough for a snack.	我零食喜欢吃酥皮点心。
9. We are glad to offer you the best chocolate mousse.	我们很高兴给你提供最佳品质的巧克力慕斯。
10. Waffle is a kind of traditional French dessert.	华夫饼是一种传统法式甜点。

舌尖聊美食

Conversation 1

Welcome to Monet's Pastry!	欢迎来莫奈的糕点店！
Hi. Could I have some souffle, and ...?	你好！来份蛋奶酥吧，还有……
Sorry, souffle is not ready yet. Could you wait for ten minutes?	对不起，蛋奶酥还没准备好呢。你可以等十分钟吗？
Oh, that's too late for me.	噢，那太晚了。
Would you like to try creme caramel? It's just hit on the tray.	要不你试试焦糖布丁吧，刚出来上架的。
Well, I'm not quite into it. I think I would like one piece of financier cake instead.	嗯，我不太喜欢焦糖布丁。我想来一块金砖蛋糕。

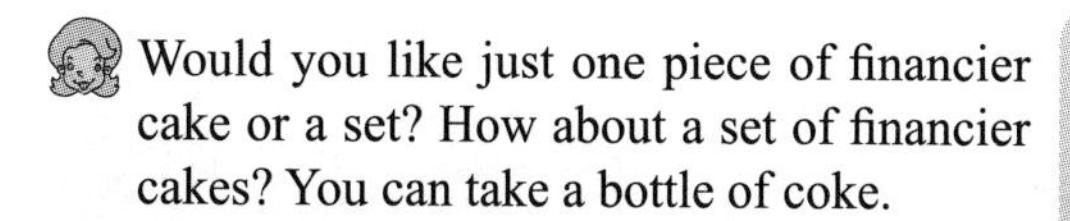

Would you like just one piece of financier cake or a set? How about a set of financier cakes? You can take a bottle of coke.

你是要一块还是一套的？要一套的话，还可以送一瓶可乐。

Thanks. I think a piece of financier cake is fine. How much is it?

谢谢。我只要一块金砖蛋糕。多少钱？

5 dollars, please.

5美元。

开胃词组

welcome to 欢迎来到……

be ready 准备，准备好

wait for 等待

be late for 对……来说太晚了

be into 喜欢，迷上

鲜美单词

pastry *n.* 糕点，油酥点心

wait *v.* 等待

minute *n.* 分，分钟

tray *n.* 盘子，托盘

piece *n.* 块，片，段

instead *adv.* 代替，反而

set *n.* （一）套，组

bottle *n.* 瓶子

coke *n.* 可乐

Conversation 2

What do you like for dessert?	你来一些什么点心?
I think I won't miss my favorite chocolate lava cake.	我想我是不会错过最喜欢的巧克力熔岩蛋糕的。
That's too fatty. And you're on diet, you remember?	那太油腻了。你在减肥呢，你忘了吗?
Oh, so I have to very careful about desserts then. Anything you recommend to me?	哦，那么我吃点心的时候得当心点。你有什么好推荐的吗?
For French suisine, I must say clafouti would be good for you, tasty and healthy.	对于法国餐，我必须说法式樱桃布丁是适合你的。
Really? Thank you. Let me have a bite of it. What about you?	真的？谢谢！那我就尝尝吧。你来点什么呢?
I like to eat chocolate fondant, a typical French dessert.	我想吃巧克力沙翁，一种特别的法式点心。
Yeah, that's a kind of soft and spongy stuff. It doesn't fit my taste.	嗯，那是一种软绵绵的东西，但不适合我的口味。

开胃词组

on diet 节食，减肥

have to 不得不

recommend to 向……推荐

be good for 对……有益，有好处

have a bite of 吃一点

鲜美单词

dessert n. 餐后甜食，甜点
favorite adj. 喜爱的，钟爱的
chocolate n. 巧克力
fatty adj. 油腻的，含脂肪的
diet v. 节食
careful adj. 仔细的，小心的
cuisine n. 烹饪，菜肴
tasty adj. 美味的，可口的
healthy adj. 健康的
fondant n. 软糖，软糖料
spongy adj. 海绵似的，海绵状的

French Desserts

法式甜品

Dessert is a course that typically comes at the end of a meal, usually consisting of sweet food but sometimes of a strongly-flavored one, such as some cheeses. The word comes from the Old French “desservir”, “to clear the table”.

甜品作为餐后的一道菜，通常包括甜食，或美味的特色食物，比如奶酪。法语有句老话“甜品就是餐桌扫尾者”。

Many dessert recipes in France call for the use of a heavy bottomed pan called a skillet. This can be crucial to the outcome of the dish, especially crepes, gaufres, (waffles) and beignets (doughnuts).

法式甜品通常用一种厚底锅叫“平底煎锅”来制作的。这种锅保证了高品质的法式甜食，比如可丽饼、华夫饼和甜甜圈。

读书笔记

Part 7

Japanese Cuisine
日本料理

Delicious Soups 美味汤

舌尖美食词汇

kenchinjiru 松肉汤	**suimono** 清汤
miso soup 味噌汤	**goma-ae** 芝麻拌菜
kinoko soup 蘑菇汤	**ton-jiru** 猪肉味噌汤
dashi 鱼汤	**kakitama-jiru** 蛋花汤
clam soup 蛤蜊汤	**niboshi dashi** 小鱼干汤
zoni 杂煮	

舌尖美食句

1. Kenchinjiru is made with miso, or with soy sauce and salt.
   松肉汤是用味噌汤、酱油和盐做成的。
2. Suimono should be served hot and be sipped directly from the bow.
   清汤应该趁热食用，一饮而尽。
3. Can you give me some suggestions for making miso soup?
   你能给我一点做味噌汤的建议吗？
4. Would you like to order a goma-ae?
   你想点芝麻拌菜吗？

5. She's good at making kinoko soup.	她擅长做蘑菇汤。
6. Ton-jiru is a rich and hearty pork miso soup.	猪肉味噌汤是一种美味的猪肉味噌汤。
7. Dashi forms the base for almost all of Japanese cooking.	日本鱼汤是日本各道菜肴会用到的高汤。
8. Japanese egg drop soup is called kakitama-jiru.	日本的鸡蛋汤被称为日式蛋花汤。
9. Clam soup is clear soup with hamaguri clam.	蛤蜊汤是一种文蛤肉清汤。
10. Niboshi dashi is Japanese fish soup stock made from niboshi.	小鱼干汤是日本一种沙丁鱼做的汤。
11. It's a Japanese tradition to eat zoni on New Year's holiday.	新年的时候吃年糕汤是日本的老传统。

舌尖聊美食

Conversation 1

Hey, did you do something for fun on the weekend?	嘿，周末有什么有趣的事吗？
Definitely! I made a Japanese soup, called miso soup, at home with my mom.	当然啦！我和我妈妈在家做了些日式汤——味噌汤。
Wow, how was it going?	哇，结果怎么样？

Not bad. It's easy to prepare it. Miso soup bases on dashi, a kind of stock, put whatever seasonal vegetables, mushrooms, tofu, meat or seafood at hand into the dashi, then add miso to the mix and serve immediately.

不错啊，很容易做的。味噌汤是用鱼汤做的，一种高汤。放各种手头上的时令蔬菜、蘑菇、豆腐、肉或海鲜到原汤里面煮，然后添加味噌，这样就可以啦。

Wow, I never thought you have a talent for cooking. Did your family like it?

哇，我未曾想到你很有厨艺天赋。你家人喜欢你做的汤吗?

Yes. They loved its salty and rich flavor, but my sister said it would be better if I added some more carrots.

是的，他们非常喜欢那咸咸的新鲜味道，但是我的姐姐说如果再放一点胡萝卜就更好了。

Come on That's not a big deal. Keep up with your miso soup.

没事，不是什么大问题啦！你的味噌汤要做得更好啊！

开胃词组

on the weekend 周末

base on 在……基础上

at hand 手边，手头上

have a talent for 在……方面有天赋

keep up with 加油，跟上，不落后

鲜美单词

fun *n.* 乐趣，趣事

definitely *int.* 当然

miso *n.* 味噌，日本豆面酱

dashi *n.* 鱼汤

stock *n.* 高汤，原汤

seasonal *adj.* 季节的，季节性的

mushroom *n.* 蘑菇

talent *n.* 天资，才能

carrot *n.* 胡萝卜

Conversation 2

Where should we have dinner?	我们去哪吃晚饭呢？
I'm not sure. What kind of food do you like?	我不知道啊。你想吃什么？
I would like to eat Japanese food tonight. I heard Japanese food is very special.	我今晚想吃日本料理。听说日本料理很特别啊。
Well, I know about a Japanese restaurant called Miki here. It's excellent. They have fresh and tasty suimono soup.	嗯，我知道这里有个叫“三木”的日本餐厅。很棒的。他们有新鲜美味的清汤。
What kind of soup is it?	这是一种什么汤？
Suimono refers to clear soup. Its name means “things to sip”. The flavors are so delicate and subtle that it is easy to go overboard with flavoring.	清汤是指一种清淡的汤品。它的名字意思就是“吸的食物”。它味道精美绝妙，添加调味品就很容易变味。
It must be a good appetizer.	那一定是一道好的开胃菜。
Not exactly. It is usually served at the end of a meal.	不是的，它一般都在餐后喝。
Oh, very strange. Anyway, let's have dinner in this restaurant.	哦，挺奇怪的。不管怎样，我们去这家餐厅吃晚饭吧。
OK, let's go!	好的，走吧。

开胃词组

have dinner 吃晚饭

know about 知道，了解

refer to 提及，涉及

go overboard 过分，过头

at the end of 在……结尾

鲜美单词

dinner *n.* 正餐，晚餐

special *adj.* 特别的，非同寻常的

restaurant *n.* 饭店，餐馆

excellent *adj.* 卓越的，杰出的

fresh *adj.* 新鲜的

refer *v.* 涉及，提及

delicate *adj.* 精美的，雅致的

subtle *adj.* 微妙的，敏感的

overboard *adv.* 越过船舷坠入水中，做得过头，过分

flavoring *n.* 调味品，调味料

Japanese soup ranges from a clear broth to thick stews filled with vegetables and noodles. The basis of most Japanese soups is the dashi which is a stock prepared from green kelp and fish along with shiitake mushrooms which provide the flavor for most of the soups. The soup is usually served in tiny decorated bowls which is placed on the right hand side of the diner. Tradition demands that it should be opposite to the rice bowl. Unlike the Westerners, the Japanese prefer to serve soup at the end of a meal along with rice and pickles.

日本汤品包括清肉汤和有蔬菜或面条的炖肉汤。日本汤品大多数以鱼汤打底，这是一种用绿色海带、鱼肉和鲜菇炖制的高汤，作为其他汤品的调味原汤。日本汤一般用精致碗盛放，并且放置于就餐者右手边。根据日本传统，汤碗要与米饭碗相对而放。与西方人喝汤的顺序不同，日本人喜欢用餐快结束时，喝汤，吃米饭和咸菜。

The miso or fermented soybean paste is another favorite flavoring agent which is added to the vegetable, fish or meat broth just before serving. It is usually consumed during breakfast in order to begin the day in a nutritious way. More elaborate variations of the miso soup may also be served for lunch or dinner.

日本的豆面酱是日本人最喜爱的调味品，可作为蔬菜类、鱼肉和肉类菜肴的调味品。早餐时，日本人也喜欢加一点豆面酱开始每一天健康营养的早餐，供午餐和晚餐食用的豆面酱品种就更多了。

读书笔记

Amazing Sushi 寿司诱惑

UNIT 02

舌尖美食词汇

inarizushi 油炸豆腐寿司，稻荷寿司	**cucumber rolls** 黄瓜卷
nigirizushi 手握寿司	**kappamaki** 黄瓜寿司
gunkanmaki 军舰卷	**temakizushi** 手卷寿司
chirashizushi 什锦寿司	**oshizushi** 模压寿司（大阪式寿司）
norimaki 紫菜卷寿司	**narezushi** 鱼饭寿司
negitoromaki 鲔鱼葱花卷	**rainbow sushi** 彩虹卷

舌尖美食句

1. I think our inarizushi is going bad.
我觉得我们的油炸豆腐寿司要变质了。

2. Cucumber rolls are called kappamaki in Japan.
在日本，黄瓜卷也被称为黄瓜寿司。

3. Nigirizushi has small rice balls with fish, shellfish on top.
手握寿司是一小块米饭团，上面有鱼肉、蚌肉。

4. Gunkanmaki is then filled with rice and fish eggs or seafood in a seaweed cup.	军舰卷在海藻杯里填上米饭和鱼籽或其他海鲜。
5. Would you like some temakizushi?	你要来点手卷寿司吗？
6. Inarizushi is a pouch of fried tofu typically filled with sushi rice.	稻荷寿司就是在油炸豆腐里面填上寿司饭团。
7. Chirashizushi is served on plates or bowls with colorful toppings.	吃什锦寿司的时候，在盘子或碗里的寿司米饭放上七彩斑斓的配料。
8. Can you help me press the sushi rice into the wooden box to make oshizushi?	你可以帮我把寿司米饭挤进木盒子里，来做模压寿司（大阪式寿司）吗？
9. Norimaki is the highlight of this dinner.	紫菜卷是这顿饭的亮点。
10. I've been thinking of trying to make my own narezushi.	我一直在考虑做自己的鱼饭寿司。
11. Negitoromaki tastes good with scallion and chopped tuna inside.	鲔鱼葱花卷尝起来味道不错，里面有葱花和碎金枪鱼肉。

舌尖聊美食

Conversation 1

Wow, is this sushi? How wonderful they are. They have red, yellow, black, and green… so beautiful feast.	哇！这是寿司吗？它们真漂亮！有红色的，黄色的，黑色的，绿色的……真是华丽的盛宴。
Yes. They're so called rainbow sushi. Colorful, isn't it?	是的，它们就是所谓的彩虹卷，五彩斑斓的，是吧？

Yes. I think they're nutritious as well. You see, it has a nice variety of fish in it.	是啊，我觉得肯定也很有营养。你看，里面有各种鱼肉。
Wow, tuna, salmon, halibut and yellowtail.	哇，有金枪鱼，三文鱼和大比目鱼，还有黄尾鲱鱼。
Rainbow roll provides you different tastes.	彩虹卷里有各种不同的味道。
Next time we'll decorate our rainbow rolls in a different way. Shall we?	下次我们做彩虹卷的时候，也用不同的方式来制作，好吗？
That's good idea.	好主意。

开胃词组

so called 所谓的

as well 也，同时

a variety of 多种的，各式各样的

next time 下次

鲜美单词

sushi *n.* 寿司

wonderful *adj.* 极好的，精彩的

feast *n.* 盛宴，大餐

rainbow *n.* 彩虹

colorful *adj.* 艳丽的，多彩的

nutritious *adj.* 有营养的

variety *n.* 各式各样，多种多样

tuna *n.* 金枪鱼，鲔鱼

salmon *n.* 三文鱼，鲑鱼

halibut *n.* 大比目鱼

yellowtail *n.* 黄尾鲱鱼

decorate *v.* 装饰

Conversation 2

What kind of foods do you like?	你想吃什么？
Well, I have eaten a lot of sushi recently.	我最近喜欢吃寿司。
Really? Is sushi a kind of sashimi?	真的吗？寿司是一种刺身吧？
No, they're totally different. Sushi refers to a dish containing rice which has been prepared with sushi vinegar. And sashimi is a kind of fresh raw meat or raw fish sliced into thin pieces. There are many different types of sushi.	不是的，寿司和刺身完全不一样。寿司是用寿司醋调味过的饭团。而刺身是生肉片或生的鱼片。寿司分为不同的种类。
OK. What's your favorite sushi?	嗯，你最喜欢哪种寿司呢？
There're nigirizushi, makizushi, inarizushi, chirashi-zushi. My favorite sushi is Futomaki. It belongs to the type of makizushi. Futomaki is a large roll piece, usually with nori outside, and cucumber strips and tiny fish roe or chopped tuna for fillings.	寿司分为手握寿司、卷寿司、稻荷寿司、什锦寿司。我最喜欢的寿司是太卷寿司，属于卷寿司的一种。太卷寿司是一大卷的饭团，外面包着海苔，里面有黄瓜条、鱼籽或金枪鱼肉末。
Oh, wow! It sounds delicious.	哦，哇！听起来很美味啊。
Yeah, it's great.	是的，很棒！
How much do they cost?	多少钱呀？

Well, it's 130 yen for one plate. They're very cheep in Japan.

每盘130日元，在日本很便宜的。

Yes, they are.

是的，是很便宜。

开胃词组

refer to 涉及，提及

raw fish 生鱼片

slice into 切开，切成

belong to 属于，归于

鲜美单词

recently *adv.* 最近地，近来地

vinegar *n.* 醋

sashimi *n.* 生鱼片

raw *adj.* 生的，未熟的

slice *v.* 切成薄片

nori *n.* 海苔，紫菜

strip *n.* 条形，条状

chop *v.* 砍，剁

plate *n.* 盘，盆

舌尖美食文化

Sushi is vinegar rice topped with other ingredients, such as fish and meat. The variety in sushi arises from the different fillings and toppings, condiments, and the way these ingredients are put together. Combined with hand-formed clumps of rice, it is called nigirizushi. Sushi served rolled inside or around nori, while dried and pressed sheet layers of seaweed or nori is makizushi. Toppings stuffed into a small pouch of fried tofu is inarizushi. Toppings served scattered over a bowl of sushi rice is called chirashi-zushi.

寿司的主要材料是用醋调味的米饭，再加上鱼肉等作为配料。种类丰富的寿司源于繁多的馅料、配料和佐料，以及将这些材料混合在一起的做法。将寿司饭捏成团状的叫手握寿司。在海藻或海苔上铺上寿司饭并加入不同的食材卷成长条的称为卷寿司。将寿司饭填满小块油炸豆腐的称为稻荷寿司。将各种食材散铺在寿司饭上层的则称为什锦寿司。

读书笔记

Japanese Main Dishes 日式主菜

舌尖美食词汇

shabu shabu 日式火锅	**tuna teriyaki** 照烧金枪鱼
nabemono 火锅	**kakuni** 烧肉块
yosenabe 什锦火锅	**chicken teriyaki** 照烧鸡肉
Kobe steak 神户牛排	**tofu teriyaki** 照烧豆腐
yakizakana 烤鱼	**tempura** 天妇罗
omusoba 蛋包面	**Tataki** 拍松

舌尖美食句

1. Shabu shabu is a Japanese variant of hot pot.
日式涮涮锅是一种日本火锅。

2. I'm really into tuna teriyaki.
我非常喜欢吃照烧金枪鱼。

3. There is nothing more delicious than nabemono.
没有什么能比得上美味的火锅了。

4. Kakuni is a Japanese braised pork dish meaning "square simmered".
烧肉块是一道日式猪肉炖菜，意思是“煨炖猪肉块”。

5. Yosenabe is a popular Japanese hot pot.	什锦火锅是很受欢迎的日式火锅。
6. Chicken teriyaki is one of my favorite dishes.	照烧鸡肉是我最喜爱的菜肴之一。
7. I can't resist the temptation of Kobe steak.	我难以抵挡神户牛排的诱惑。
8. I'm going to prepare an original tofu teriyaki.	我打算做一个地道的照烧豆腐。
9. Yakizakana is a great special dish in Japan.	烤鱼是日本别有风味的一道菜。
10. Tempura is a popular Japanese food as well-known as sushi.	天妇罗是一种与寿司齐名的日本菜式。
11. Omusoba is an easy dish, remindful of what you would get in a Japanese household.	蛋包面是很容易做的一道菜，是令人难以忘怀的日本家常菜。
12. Tataki is a method of preparing fish or meat in Japanese cuisine.	拍松是日式料理中鱼肉类菜肴的一种烹饪方法。

舌尖聊美食

Conversation 1

What are we going to eat for dinner?	我们今晚吃什么？
How about yosenabe?	吃海鲜火锅怎么样？
Good. That's a good choice for this cold winter.	好，冬天吃火锅真是一个好主意。

Right. Everything is boiling in a hot pot, such as seafood and vegetable. It's going to be fun and exciting.	没错。所有东西都放到一个热锅里煮，海鲜和蔬菜。好玩又开心。
Great! How about having dashi soup for the soup base?	真棒！用鱼汤做锅底怎么样？
It sounds good. And what ingredients we shall add to it?	不错。我们应该放什么菜？
Hmm… I think clams, salmon slice, cabbage, leek, carrot, mushroom…and you?	嗯……蛤蜊、三文鱼片、大白菜、韭菜、胡萝卜、蘑菇……你觉得呢？
Tofu is a must.	豆腐是必点的哦。
Yes, and noodles!	好的，还有面条！
That would be nice. Let's have a marvelous dinner!	那很好。我们来享用丰盛的晚餐吧！

开胃词组

hot pot 火锅

soup base 锅底

add to 添加，增加

have dinner 吃晚饭

鲜美单词

choice *n.* 选择，选择权

boil *v.* 用开水煮，用沸水煮

seafood *n.* 海鲜

vegetable *n.* 蔬菜

exciting *adj.* 令人兴奋的

base *n.* 主要成分，主料

ingredient *n.* 原料

clam *n.* 蛤蜊，蚌类

cabbage *n.* 卷心菜，洋白菜

marvelous *adj.* 非凡的，惊人的

Conversation 2

What are you going to cook today?	你今天做什么菜？
I am going to do a Japanese cuisine, called Yakizakana, grilled dried herring.	我正要做日本菜，叫烤鱼，烤鲱鱼。
I thought you like to make sashimi.	我以为你要做刺身呢。
Well, when it comes to fish, you might think of sashimi in Japan. But Japanese people also like to grill fish.	嗯，在日本每当说到鱼，你也许会想起刺身。但是日本人也喜欢烤鱼。
Oh, really? OK. How are you going to do Yakizakana?	噢，真的吗？好吧!你打算如何做烤鱼呢？
There are many kinds of drying processes used for fish here. Dried overnight with just a bit of salt is very common and tasty. Dried with varying degrees of saltiness: sweet, middle and very salty is also common. This herring is dried with some salt and its own eggs filling the entrails cavity. This herring wasn't dried rock hard, but was semi-dried and soft and juicy after cooking.	这里的烤鱼有许多种，加一点点盐，晾干一个晚上，这种烤鱼很受欢迎，很美味。根据不同咸度，有甜的、中等咸度的、咸味十足的等，也很受欢迎。我这个烤鲱鱼，加一点点盐，胸腔内还带有鱼籽。这份烤鱼不烤到肉质干硬的程度，只烤到有点干即可，做好之后肉质柔软多汁。
How does it taste?	吃起来味道怎么样呢？

Salty, but not too salty, just the right amount to contrast well with rice. I used the leftovers for chazuke, tea poured over flavored rice, the chopped herring adding the flavor.

吃起来有咸味，但不是很咸，味道正好与米饭相衬。我通常用剩下的鱼肉做茶渍饭，用茶水倒进香甜的米饭，把切碎的鲱鱼肉放进去，味道十足啊！

Alright, it sounds yummy!

嗯，听起来很好吃！

开胃词组

think of 想出，想起

a bit of 一点儿

pour over 倒入

鲜美单词

cook *v.* 烹调，做菜

grill *v.* 烧烤

herring *n.* 鲱鱼

sashimi *n.* 生鱼片

process *n.* 加工，处理过程

overnight *adv.* 在晚上，在夜里

entrails *n.* 内脏

cavity *n.* 腔，空腔

amount *n.* 量，数量

leftover *adj.* 剩余的，未用完的

Sashimi is fresh, raw fish that is sliced very thinly and served uncooked. As traditional Japanese dish, it is usually served with daikon radish, pickled ginger, wasabi and soy sauce as the first course in a meal.

刺身是切成薄片的，生吃的新鲜鱼肉片。作为日本料理非常传统的一道凉菜，通常与日本白萝卜、生姜片、芥末、酱油一起食用。

As sashimi is eaten raw, the fish must be best quality, and it should be as fresh as possible. Some Japanese restaurants keep the fish alive in tanks right up to the minute they prepare it. Highly skilled chefs train for years to perfect the art of slicing the fish, according to variety, to maximise enjoyment. Having said that, it is possible to make sashimi at home, so long as the fish is super-fresh, and it's handled properly.

刺身是生吃的，鱼肉质量非常讲究，要尽可能新鲜,日本一些餐馆把活鱼放在容器里养到做菜的时候。厨艺精湛的厨师需要几年的训练，才能做出最完美的、不同种类的生鱼片。当然，在家也能做生鱼片，但要求鱼必须是非常新鲜，正确处理和烹制。

读书笔记

Healthy Noodles 健康面条

舌尖美食词汇

ramen 拉面	**yakisoba** 炒面
zaru soba 笼屉荞麦面	**tempura soba** 天妇罗荞麦面
curry udon 咖喱乌冬面	**shirataki** 魔芋丝
somen 素面	**miso ramen** 日式味噌拉面
tonkotsu ramen 猪骨汤拉面	**yakiudon** 炒乌冬面
nabeyaki udon 砂锅面条	

舌尖美食句

1. Ramen are made with wheat flour, water, salt and kansui.	拉面是用白面、水、盐和碱水做的。
2. Would you like a bowl of yakisoba?	你想来一碗炒面吗?
3. Zaru soba is served with dipping sauce.	笼屉荞麦面要蘸着酱汁来吃。
4. As I rarely make tempura at home, I like ordering tempura soba when I'm out.	因为我在家不经常做天妇罗，所以在外面吃饭的时候，我经常点天妇罗乌冬面。

5. Do you know how to make curry udon?	你知道怎么做咖喱乌冬面吗？
6. Shirataki are chewy or rubbery.	魔芋丝耐嚼，有韧性。
7. During the summer Japanese consume chilled sōmen to stay cool.	在夏季，日本人吃素面来消暑。
8. Lots of vegetables can be added in miso ramen.	日式味噌面里面可以放很多蔬菜。
9. I love to eat this tonkotsu ramen.	我喜欢吃这种猪骨汤拉面。
10. Yakiudon is popular and usually sold as street food in Japan.	炒乌冬面在日本很受欢迎，通常在街头可以买到。
11. I don't like nabeyaki udon at all.	我一点也不喜欢砂锅面条。

舌尖聊美食

Conversation 1

Hi, Lucy! You're back. How is your visit to Japan?	嗨，露西！你回来了？你日本之行怎么样？
Well, that's fantastic! You know what, I love Japanese food, especially the delicious noodles.	非常棒！知道吗，我喜欢日本食物，尤其那美味的面条。
Great! Are Japanese noodles different from Italian pasta or Chinese noodles?	好！日本面条跟意大利面或中国面条不同吗？
Oh, yes. They're flavorful and healthy based on all kinds of sauce.	喔，是的。日本面条用各种各样的酱料，美味又健康。
What type of noodles do you like best?	你最喜欢哪种面条？

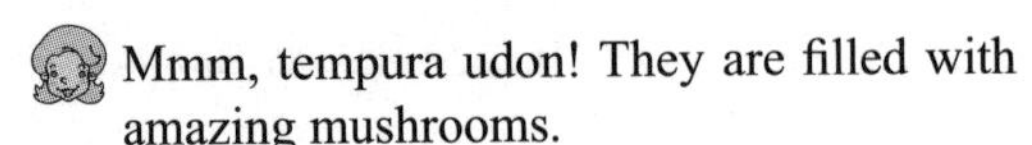
Mmm, tempura udon! They are filled with amazing mushrooms.

嗯……天妇罗乌冬面！里面有非常好吃的蘑菇。

Wow, you're a fan of udon.

哇，看来你非常喜欢吃乌冬面啊。

Absolutely. How about coming to my house this weekend? And I prepare tempura udon for you.

当然！周末来我家，怎样？我做天妇罗乌冬面给你吃。

Ha! That's really nice.

哈！很好啊。

开胃词组

be different from 不同于……，异于……

base on 基于，在……基础之上

fill with 填满，使充满

a fan of ……的粉丝

鲜美单词

visit n. 参观，访问

fantastic adj. 极好的，了不起的

especially adv. 尤其，特别地

delicious adj. 美味的，可口的

pasta n. 意大利面

flavorful adj. 可口的

healthy adj. 健康的

sauce n. 酱，调味汁

tempura n. 天妇罗

mushroom n. 蘑菇

udon n. 乌冬面

Conversation 2

Mr. and Mrs. Takeda, thank you very much for inviting me to come.

武田先生，武田夫人，非常感谢你们邀请我。

We're glad to have dinner with you. Please help yourself with shoyu ramen, the most popular needles in Japan.

我们很高兴与您一起共进晚餐。请来点酱油拉面吧，在日本最受欢迎的面条。

Thank you. Wow, I love the shoyu and the kombu dashi soup stock in it.

谢谢！哇，我喜欢里面的酱油和紫菜清汤。

Yes. This noodle dish is prepared with lots of soy sauce, which is known as shoyu in chaste Japanese. Chukamen noodles, garlic, fresh ginger, sesame oil, chicken soup stock, kombu dashi soup stock, sake, salt, sugar, and soy sauce are used to make this dish.

是的，这种面条用酱油烹制，这种酱油也叫日本酱油。这道菜是用中式面条、大蒜、生姜、芝麻油、鸡汤、紫菜清汤、米酒、盐、糖、酱油等来做的。

I find Japanese noodles are all delicious and healthy. They aren't as heavy as pasta, usually use vegetable or fish to prepare, very healthy.

我觉得日本面条都很美味又健康。没有意大利面口味那么重，通常使用蔬菜、鱼等材料来做，非常健康。

That's true. We have different type of noodles, like ramen, soba, shirataki, udon, somen. Most noodle dishes are prepared with ingredients such as dashi, shoyu, mirin eggs, chicken, fish and vegetables are used in many variants.

没错。我们有各种不同的面条，比如拉面、荞麦面、魔芋丝、乌冬面、素面。大多数面食都用清汤、清酱油、料酒等来烹制，通常有鸡蛋、鸡肉、鱼肉、蔬菜等。

Japanese noodles are great!

日本面非常棒！

开胃词组

help yourself 随便吃

be known as 被认为是，号称

such as 例如

鲜美单词

invite *v.* 邀请，请求

kombu *n.* 海带

stock *n.* 原汤，高汤

shoyu *n.* 酱油

chaste *adj.* 童真的，纯洁的

sesame *n.* 芝麻

sake *n.* 日本米酒

variant *n.* 变化

舌尖美食文化

Japanese noodles have gain major popularity outside of Japan, especially with the recent boom of Sushi in the last 5 years. Much like rice, noodles are a staple and often used as a focal point in many Japanese dishes. In Japan's towns and cities, noodle shops can be found at every corner.

日本面条越来越受外国人欢迎，尤其近5年来掀起寿司风潮之后。跟米饭一样，面条也是日本的主食之一，是日本菜的焦点。不管在日本的城市还是小镇，面馆遍布大街小巷。

Contrary to popular belief, Japanese noodles can also be served cold with variety of dipping dishes, but more frequently used in hot and cold soups, accompanied by healthy ingredients such as mushrooms, meats, fish, kelp and so on.

与大家想象的完全不一样，日本面食也有蘸酱吃的冷面，但大部分都是带有热汤或冷汤的面食，里面还有健康的食物，比如蘑菇、肉、鱼、海带等等。

There are a variety of Japanese noodles such as ramen, soba, udon, somen and shirataki. Each style of noodles has its own unique characteristics, suitable for a variety of different dishes.

日本有许多种面条，有拉面、荞麦面、乌冬面、素面和魔芋丝。每种面均有自己独特的风味作为菜肴特色。

读书笔记

读书笔记

Part 8

South Korean Cuisine
韩式料理

UNIT 01

Kimchi 韩国泡菜

舌尖美食词汇

baek kimchi 未加辣椒粉的泡菜	**yeolmu kimchi** 水萝卜泡菜
dongchimi 冬泡菜（用整个萝卜腌的泡菜）	**baechu kimchi** 辣白菜泡菜
bossam kimchi 包卷泡菜	**chonggak kimchi** 萝卜块泡菜
baechu geotjeori 即食（鲜白菜）泡菜	**oisobagi** 填黄瓜泡菜
pa kimchi 大葱泡菜	**watery kimchi** 水泡菜

舌尖美食句

1. Baek kimchi isn't spicy at all.

未加辣椒粉的泡菜一点也不辣。

2. Yeolmu kimchi is popular in the summer.

水萝卜泡菜在夏天很受欢迎。

3. Try this dongchimi, very sweet.

尝尝这个鲜萝卜泡菜，非常甜的。

4. Is the baechu kimchi too spicy for you?	辣白菜泡菜对你来说太辣了吗？
5. Bossam kimchi is a style of kimchi that has a lot of different things wrapped inside.	包卷泡菜是一种里面包有许多东西的泡菜。
6. I would like to introduce you this chonggak kimchi.	我想给你介绍这种萝卜泡菜。
7. There is a Korean dish for fresh kimchi called baechu geotjeori.	有一种新鲜的未发酵的韩国泡菜叫即食泡菜。
8. Oisobagi is known as stuffed cucumber kimchi.	黄瓜泡菜是一种在黄瓜里填上配料的泡菜。
9. You can enjoy this flavorful pa kimchi.	你可以尽情享用这美味的大葱泡菜。
10. Can I get some more napa cabbage kimchi and watery kimchi?	我可以再要点大白菜泡菜和水泡菜吗？

舌尖聊美食

Conversation 1

What kind of kimchi do you like?	你喜欢哪种泡菜？
I like a type of kimchi that is called watery kimchi. Do you know how to make it?	我喜欢那种水泡菜。你知道怎么做吗？
Oh, yes, you just need to soak it into the water.	是的，你只需要把它往水里泡就可以了。
What are the main ingredients then?	主要用些什么食料呢？

This dish includes cabbage, carrot, watercress, nuts and other seasonings. Just like other kimchi dishes. The difference is that it has broth when it is served.

这道菜用大白菜、胡萝卜、水田芥、果仁和其他酱料来制作，就像其他泡菜的制作方法一样。不同的地方就是吃的时候，这种泡菜是带汤的。

Oh, I see. Is the broth tasty to drink?

哦，明白了。这种汤好喝吗？

Definitely. Many people love to drink it, or you can serve it with other cold dishes as well.

当然，许多人都喜欢喝，也可以与其他冷盘一起吃。

May I learn how to make it with you sometimes?

我跟你学习怎么制作这道泡菜，可以吗？

Yeah, how about dropping by my house this weekend? We do it together.

可以，这周末你来我家，我们一起做怎么样？

Good idea.

好主意。

开胃词组

need to 需要

cold dish 冷盘，凉菜

as well 也，还有

drop by 顺便来访，造访

鲜美单词

kimchi *n.* 韩国泡菜

type *n.* 类型，种类

soak *v.* 浸，泡

ingredient *n.* 原料，材料

cabbage *n.* 卷心菜，洋白菜

watercress *n.* 水田芥

seasoning *n.* 调味品，佐料

broth *n.* 肉汤

tasty *adj.* 美味的，可口的

serve *v.* 上菜；招待

sometimes *adv.* 有时，有时候

together *adv.* 一起，共同

Conversation 2

Hi, Kim. I heard that you're good at cooking Korean cuisine. Could you tell me something about kimchi?	你好，金先生。我听说你擅长做韩国菜。你能给我介绍一下韩国泡菜吗？
Sure! Well, Kimchi is a national Korean dish consisting of fermented chili peppers and vegetables, usually based on cabbage. It is also a nutritious dish, providing vitamins, lactic acid and minerals. Kimchi can also be preserved for a long time.	当然可以！嗯，泡菜是韩国的国菜，主要用辣椒和蔬菜，大多数用大白菜腌制而成。泡菜很有营养，为人体提供维生素、乳酸和矿物质。此外，它还能保存很长时间。
OK, so does pickles. Is there more than one type of kimchi?	没错，泡菜就是这样，泡菜不止一种吧？
Yes. There are many types of kimchi. The most common kimchi variations are baechu kimchi, baechu geotjeori, bossam kimchi, baek kimchi, dongchimi, chonggak kimchi, kkakdugi, oisobagi and pa kimchi.	是的，泡菜有很多种。最常见的有：韩式辣白菜、未发酵白菜泡菜、包卷泡菜、未加辣椒粉白菜泡菜、冬泡菜、萝卜泡菜、萝卜块泡菜、填黄瓜泡菜、大葱泡菜。
How to make this one? It looks lovely and smells good, too.	这个怎么做？它看起来很好看，闻着也很香。

Oh, this is baechu kimchi. It is the most popular winter kimchi made by stuffing, the blended stuffing materials, between the layers of salted leaves of uncut, whole cabbage.

噢，那是辣白菜泡菜。最受欢迎的冬季泡菜，整个白菜菜叶层之间放填充物，腌制而成。

OK, I see. Are there any others?

噢，我知道了。还有其他的吗？

Yes, we have yeolmu kimchi. Although they are thin and small, young summer radishes are one of the most common vegetables for kimchi during the spring and summer season.

有，我们还有水萝卜泡菜。水萝卜虽然又细又小，但是它可是春夏常用的泡菜原材料。

Wow! And this is just a selection? Which is your favourite?

哇，这只是其中的一部分吧？你最喜欢哪一种？

I love pa kimchi.The hot spicy pa kimchi, most popular in Jeolla-do, is made of medium-thick young green onions.

我喜欢大葱泡菜。热辣的大葱泡菜，在全罗道最有名，由中等大小的大葱腌制而成。

Well, all these kimchi dishes must be delicious.

嗯，这些泡菜肯定都非常好吃！

开胃词组

consist of 由……组成

chili pepper 红辣椒

a long time 很长的一段时间

be made of 用……做成

鲜美单词

national adj. 国家的，国有的

ferment v. 使发酵

nutritious adj. 有营养的

vitamin n. 维生素

lactic adj. 乳的

mineral n. 矿物，矿物质

preserve v. 保存，腌制

pickle n. 腌菜，泡菜

variation n. 变化，变动

blend v. 混合，掺杂

layer n. 层，层次

radish n. 小萝卜

舌尖美食文化

Kimchi is a national Korean dish consisting of fermented chili peppers and vegetables, usually based on cabbage. It is suspected that the name kimchi originated from shimchae (salting of vegetable) which went through some phonetic changes: shimchae — dimchae — kimchae — kimchi.

泡菜是韩国的国菜，由红辣椒和蔬菜腌制而成，通常以大白菜为主。据推测，韩国泡菜的名称从shimchae（腌制蔬菜）而来，通过语音的演变而来：shimchae变成dimchae，再变成kimchae，最后变成kimchi。

Common ingredients include Chinese cabbage, radish, garlic, red pepper, spring onion, ginger, salt and sugar. While kimchi is generally identified internationally as Chinese cabbage fermented with a mixture of red pepper, garlic, ginger and salted fish sauce, several types of kimchi exists, including regional and seasonal variations.

主要原材料包括大白菜、萝卜、红辣椒、大葱、生姜、盐和糖。因而泡菜一般是用大白菜和红辣椒、大蒜、生姜和咸鱼酱，还有其他泡菜原料腌制而成的。不同地区，不同季节的泡菜也不同。

Kimchi has been cited by Health Magazine as one of the world's five "healthiest foods", with the claim that it is rich in vitamins, aids digestion, and may even prevent cancer. The health properties of kimchi are due to a variety of factors. It is usually made with cabbage, onions and garlic, all of which have well-known health benefits. It also has active and beneficial bacterial cultures, like yogurt. Lastly, kimchi contains liberal quantities of red chili peppers which has been suggested to have health benefits as well.

泡菜被健康杂志认为是世界上五大"健康食物"之一，称泡菜富含维生素、助消化物质，还可以预防癌症。泡菜有益健康成分主要归因于各种营养要素，用来制作泡菜的大白菜、洋葱和大葱等都是非常有益健康的食材。泡菜还像酸奶一样，含有有益菌成分。泡菜里的许多红辣椒也是有益健康的调味品。

读书笔记

Main Dishes
主菜

舌尖美食词汇

maeun-tang 辣汤	**nakji-jeongol** 辣味章鱼火锅
Korean barbecued chicken 韩国烤鸡	**budae jjigae** 部队汤
samgyeopsal-gui 烤五花肉	**samgye-tang** 参鸡汤
jangeo-gui 烤鳗鱼	**galbi** 烤小牛排
Andong jjimdak 安东蒸鸡	**pajeon** 朝鲜葱饼

舌尖美食句

1.	Maeun-tang is dish of hot spicy fish soup.	辣汤是一道热辣的鱼汤。
2.	Nakji-jeongol is my favorite dish of Korean cuisine.	辣味章鱼火锅是我最喜欢的一道韩国菜。
3.	Korean barbecued chicken are doused in a spicy-sweet soy-sesame-chile marinade, then grilled and wrapped up in lettuce leaves to eat.	做韩国烤鸡，先把鸡肉放进甜辣的芝麻酱里腌制，烤熟之后用生菜叶包着吃。

4. The Korean people usually grill the entire fish with simple seasonings such as salt, soy sauce, or hot pepper sauce.	韩国人烤鱼通常只放一点调味品，比如盐、酱油或辣椒酱。
5. We used to eat budae jjigae instead of sundae.	我们常吃部队汤，而不常吃韩式香肠。
6. Don't eat too much samgyeopsal-gui, or you would gain a lot of weight.	不要吃太多烤五花肉，否则你会长胖的。
7. Samgye-tang is healthy and tasty.	参鸡汤健康又好喝。
8. Do you know how to prepare jangeo-gui?	你知道怎么做烤鳗鱼吗？
9. Galbi is usually cooked on a metal plate over charcoal in the centre of the table.	烤小牛排通常在桌子中间的炭火炉上面烤。
10. Andong Jjimdak is a spicy dish of steamed chicken with vegetables and cellophane noodles in ganjang sauce.	安东蒸鸡是一道辣菜，鸡肉与蔬菜、面条，再放点酱油一起蒸。
11. Pajeon is a kind of pancake made mostly with eggs, flour, green onion, and oysters.	朝鲜葱饼大多是用鸡蛋、面粉、大葱和牡蛎做的。

舌尖聊美食

Conversation 1

OK, Nacy. Help yourself to my homemade dishes.	好了，娜西，请随便吃我做的家常菜。
Thank you for your hospitality, Fiona. I'm very glad to have a taste of your specialties.	菲奥纳，非常感谢您的热情款待。我很高兴能尝你的拿手好菜。
Oh, that's jangeo-gui, a traditional Korean dish. Have a try, please.	哦，那是烤鳗鱼，一道传统的韩国菜。你尝一尝。

So that's the famous grilled eel? It tastes like what I had in a Korean restaurant, good looking and flavorful. How do you make it?

那么这就是那著名的烤鳗鱼了？吃起来就像我在韩国餐厅吃的一样，好看又可口。你是怎么做的？

Oh, very simple. Eels are sliced longways and the bones removed before being seasoned with sesame oil, sesame seeds, soy sauce and sugar. The strips are then broiled.

噢，非常简单的。纵向切成条，去骨，配上芝麻油、芝麻、酱油和糖。然后烤就可以了。

That's it?

就这样？

Well, it would be even better to get fresh eels when grilling them.

嗯，烤的时候，用新鲜的鱼会更好些。

Wow, it sounds interesting.

哇，听起来很有趣！

开胃词组

help yourself to 请自便，随意

have a taste of 尝一尝

have a try 尝试一下

be seasoned with 用……来调味

鲜美单词

homemade *adj.* 家庭自制的

hospitality *n.* 殷勤好客；款待

specialty *n.* 专长；特色菜

traditional *adj.* 传统的，惯例的

famous *adj.* 著名的，出名的

Korean *adj.* 朝鲜（或韩国）的，朝鲜（或韩国）人的

restaurant *n.* 饭店，餐馆

flavorful *adj.* 可口的

longways adv. 纵长地

broil v. 烤，焙

Conversation 2

Jason, you have been wondering if you have a chance to try some authentic Korean cuisine. Now, here it is. This is a traditional Korean soup.	詹森，你一直想借个机会尝尝正宗的韩国菜。那么，这道菜就是。这是一道韩国传统汤品。
What's in it?	里面有什么呢？
This is a soup with chicken and ginseng. It's called samgye-tang in Korea.	这是鸡肉人参汤，用韩语说，就是参鸡汤。
Chicken and ginseng? It sounds a bit strange. Is it healthy?	鸡肉和人参？听起来有点奇怪。健康吗？
Yes. It's specially served in summer time. This soup could make up those nutrients we lost in the daily excessive sweating.	是的。通常在夏天的时候喝。这个汤能补充因流汗过多而损失的营养成分。
Why do these ingredients have not been cut?	这些东西为什么都没有切片呢？
That's because we belive they could preserve the maximum amount of nutrients.	那是因为我们相信不切片，能最大限度地保留营养物质。
Wow, amazing. It seems a bowl of samgye-tang a day keeps the doctor away.	哇，真奇妙！一天一碗参鸡汤，就不用看医生啦！

开胃词组

wonder if 想知道是否……

summer time 夏季

make up 补足，补偿
it seems... 似乎……
keep away 使不靠近，远离

鲜美单词

wonder *v.* 想知道，想弄明白
chance *n.* 机会，机遇
authentic *adj.* 真的，正宗的
cuisine *n.* 烹饪，菜肴
ginseng *n.* 人参的，高丽参
strange *adj.* 陌生的，奇怪的
nutrient *n.* 营养物质，养分
excessive *adj.* 过度的，极度的
ingredient *n.* 原料，材料
preserve *v.* 保存，腌制
maximum *adj.* 最大值的，最大量的

舌尖美食文化

Korean cuisine involves the use a lot of garlic (a lot more than in Thai food, Italian, Spanish or Greek cuisine), a lot of red chillies, spices such as ginger, doenjang, soy sauce and gochujang (red chilli paste). The cooking oil normally used by Koreans is sesame oil. Korean cuisine includes recipes with meat, fish, vegetables, noodles and tofu. Altogether Korean cuisine is very healthy.

韩国菜一般使用很多大蒜（用量比泰国菜、意大利菜、西班牙菜

或希腊菜还多），大量的辣椒，还有生姜、黄酱、酱油和辣椒酱等调味品。韩国菜食用油一般使用芝麻油。韩国菜有肉、鱼、蔬菜、面条和豆腐。总的来讲，韩国菜是非常健康有营养的。

Korean cuisine has been affected by its geography (peninsula), climate (hot, humid summers and very cold winters), proximity to neighbors China and Japan European traders also had an impact in the cuisine with the Portuguese introduction of chili peppers to Korea in the 17th century. By the 18th century, chili peppers were already being widely used in the preparation of Korean cuisine.

韩国菜受地理位置（半岛）、气候（夏季炎热、潮湿，冬季严寒），还有近邻国家中国、日本菜式的影响。欧洲人与韩国的贸易活动也对韩国饮食产生了影响，葡萄牙人于17世纪把辣椒带到了韩国，到了18世纪，韩国菜普遍使用辣椒作为食材。

读书笔记

UNIT 03

Desserts 甜品

舌尖美食词汇

baekseolgi 白蒸糕	**hoppang** 豆沙包
yeot 饴糖	**green tea cakes** 绿茶蛋糕
kkultarae 宫廷饼	**songpyeon** 松糕
hangwa 韩果	**yaksik** 八宝饭
Korean scallion pancake 韩国葱饼	**candied ginger** 姜糖
injeolmi 切糕	**sweet custard bread** 香甜乳蛋包

舌尖美食句

1. Baekseolgi often appears on the tables of babies' birthdays and other celebrations.
白蒸糕通常出现于儿童生日庆祝会或其他庆祝会。

2. Hoppang is a popular hot snack commonly found in Korea especially during the winter season.
豆沙包在韩国是一种受欢迎的点心，尤其在冬季的时候。

3. Yeot is Korean traditional confectionery.
饴糖是一种韩国传统甜食。

4. I would like two green tea cakes to go.
我要两份绿茶蛋糕, 带走。

5. Kkultarae is tasty with a sweet nut filling.	带香甜果仁的宫廷饼很美味。
6. Koreans usually eat Songpyeon during the Korean autumn harvest festival, Chuseok.	韩国人通常在秋日丰收季节，即中秋节，吃松糕。
7. Hangwa is a kind of traditional cookie in Korea.	韩果是韩国的一种传统饼干。
8. It's not that easy to make yaksik.	八宝饭不是那么容易做的。
9. I don't like the heavy taste of Korean scallion pancake.	我不喜欢韩国葱饼浓重的味道。
10. Where can I get some candied ginger?	哪里可以买到姜糖?
11. Injeolmi is easily digested and nutritious.	切糕容易消化又有营养。
12. Would you like some sweet custard bread?	你想来点香甜乳蛋包吗?

舌尖聊美食

Conversation 1

Wow, these look amazing. What are these?	哇，这些看起来很棒啊！这些是什么?
They're tteok, a traditional Korean dessert. They're very popular in Korea.	这是一些打糕，韩国传统的点心。它们在韩国很受欢迎的。
They look like Japanese daifuku. But they have different colors, white, pink, green and yellow.	它们看起来就像日本的大福糕，但是它们有不同的颜色，白色的、粉色的、绿色的，还有黄色的。

Well, these brightly colored tidbits also have a flavor all their own.	嗯，这些不同颜色的打糕味道也不一样。
Really? Is there any stuffing inside?	真的吗？里面包有馅儿吗？
Yes, these are Baram tteok, a kind of rice cake. They usually contain red bean paste inside.	是的，这些是巴拉姆糕，一种米糕。它们里面含有红豆沙。
It sounds delicious.	听起来很美味啊！
By the way, they are often half-moon-shaped rice cakes that contain sweet red bean paste. They have become a popular symbol of traditional Korean culture.	顺便说一下，这种米糕通常做成半月形的，里面包着红豆沙，成为韩国的传统文化象征的食物。
Oh, I see.	哦，明白了。

开胃词组

be popular in 在……流行，受欢迎

look like 看起来似乎……

rice cake 年糕，米糕

by the way 顺便提一下，顺便说

鲜美单词

dessert n. 甜点，餐后甜点

brightly adv. 明亮地，鲜明地

tidbit n. 少量的美食

flavor n. 味道，香味

paste n. 面团，糨糊

delicious adj. 美味的，可口的

symbol n. 象征，标志

culture n. 文化

Conversation 2

I would like to introduce you a traditional Korean sweet dish, called yaksik. It can be a snack or dessert.	我想要给你介绍一个传统韩国甜品，叫八宝饭。它可做零食或点心。
OK. What's the ingredient anyway?	哦，什么做的？
It is made with sweet rice, nuts and jujubes, a kind of Korean dates. And yaksik got its name due to the use of honey in its ingredients.	它是用甜米、果仁和韩国枣做的。八宝饭得名于它的原材料蜂蜜。
Yes? So the name of yaksik means sweet?	真的吗？八宝饭的名字意思是甜吗？
Kind of, though. Actually, "Yak" means medicine, and "Sik" means food. It is said that honey was commonly called as yak, meaning medicine.	有点那个意思。实际上，“Yak”是指“药”，“Sik”是指“食物”。据说蜂蜜被认为是一种药，叫“yak”。
How intersting it is.	真有意思。
Many Koreans eat yaksik as a healthy alternative to cookies and chips. It's often served at weddings and parties.	许多韩国人喜欢吃健康的八宝饭，而不吃饼干或薯条。八宝饭通常是婚礼和派对必备食物。
I think I would like to try some.	我觉得我想尝一下。
OK, let's go and buy some. They are available in every pastry shop.	好，我们去买点八宝饭。各糕点店都有卖的。

开胃词组

due to 因为，由于；归因于

kind of 有点，稍微

be available 可获得，可拿到

鲜美单词

introduce *v.* 介绍，引进

snack *n.* 快餐，点心

jujube *n.* 枣，枣味软糖

honey *n.* 蜂蜜

medicine *n.* 医药，药品

commonly *adv.* 通常地，一般地

interesting *adj.* 有趣的，令人感兴趣的

alternative *adj.* 替代的，备选的

available *adj.* 可获得的，可用的

pastry *n.* 糕点，点心

舌尖美食文化

Korean desserts can also be very sweet. They may include cookies, rice cakes, ice creams as well as some sweet pastries. Songpyeon, a famous rice cake, is served during a festival at autumn called Chuseok. This rice cake is decorated with seeds and nuts. Another rice cake called Ddok, which is typically formed into different sizes and shapes and are also soft and chewy.

韩国甜点非常甜，包括饼干、米糕、冰淇淋，还有其他的面包点心。松饼是韩国著名的米糕，一般在秋收季节的时候吃，这个节日也叫中秋节。松糕外皮用果子或果仁点缀。另一种米糕，也叫年糕，做成不同的大小和形状，柔软而耐嚼。

Other common desserts in Korea are rolls such as ho-ddok, which comes with a cinnamon and honey filling in the inside. Pot-bingsu is the Korean version of the ice cream, which is made of mainly sugar sauce, red beans, crushed ices and fruits. A dough composed of sweet red beans called boong-aw bbang is another popular dessert.

韩国其他受欢迎的甜品有各种卷，例如ho-ddok，里面含有肉桂和蜂蜜。红豆刨冰是韩式冰淇淋的一种，主要用糖浆、红豆、碎冰和水果做的。红豆馅面包也是在韩国受欢迎的甜品。

读书笔记

Korean Soups 韩式汤

舌尖美食词汇

ox bone soup 牛骨汤	**naengguk** 凉汤
gomguk 牛肉汤	**kimchiguk** 泡菜汤
bean sprout soup 豆芽汤	**bosin-tang** 补身汤
tojangguk 大酱汤	**altang** 香辣鱼子汤
chueotang 泥鳅汤	**seaweed soup** 海带汤
tteokguk 年糕汤	**mandutguk** 饺子汤

舌尖美食句

1.	Ox bone soup is very popular all year round in Korea.	牛骨汤在韩国常年受欢迎。
2.	Naengguk refers to all kinds of cold soups in Korean cuisine.	凉汤是韩式菜里的各种冷汤。
3.	Shall we start with gomguk?	我们首先喝牛肉汤吗？
4.	Do you prefer kimchiguk to bean sprout soup?	与豆芽汤相比，你更喜欢泡菜汤吗？

5. Bosin-tang is a kind of dog meat soup.	补身汤是一种狗肉汤。
6. Tojangguk is usually seasoned with doen-jang.	大酱汤通常用黄酱调味。
7. Altang is too spicy for me.	香辣鱼子汤对我来说太辣了。
8. Where I can get chueotang, a spicy Korean mudfish soup?	我在哪可以喝到泥鳅汤，一种韩国鱼汤?
9. Seaweed soup is commonly eaten in South Korea, especially after child birth.	在韩国海带汤通常在刚生完孩子的时候喝。
10. Tteokguk is a traditional rice cake soup eaten during the celebration of the Korean New Year.	年糕汤是一种传统的韩式米糕汤，通常在庆祝新年的时候喝。
11. Mandutguk is basically Korean dumpling soup.	饺子汤是一种韩式饺子汤。

舌尖聊美食

Conversation 1

Simon, would you like a bowl of soup to go with your rice?	西蒙，来一碗汤泡饭吗?
Rice with soup? It sounds interesting. I never try this before.	汤泡饭真有意思，我从来没有这么吃过。
In South Korea, most of soup comes with rice or they prepare the soup with rice inside.	在韩国，大多数时候，汤要与米饭一起吃的，或者他们做汤的时候，里面已经放米饭了。

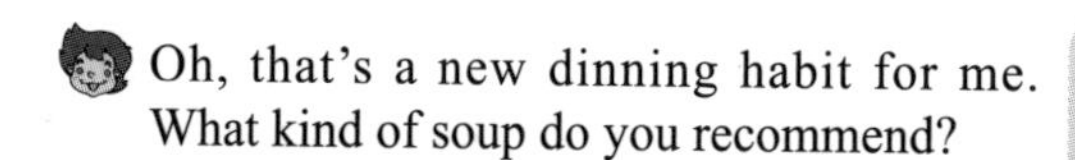

Oh, that's a new dinning habit for me. What kind of soup do you recommend?

哦，这对我来说是一种新鲜的吃法。你推荐什么汤呢？

How about this one? Ox bone soup, a milky white soup made of long-simmered ox leg bones. The aroma of ox bone soup is rich and meaty, and the broth sticks to the roof of your mouth.

这个牛骨汤怎么样？长时间煨炖的牛骨汤，奶白色的。牛骨汤的芳香四溢，让你回味无穷。

It's cold here. I think soups can warm our body up during the winter time.

这里很冷。我觉得在这样的冬天喝点汤也能暖和身体。

You can serve it with kimchi as well, or you can make it with radish and add sliced brisket meat and noodles.

你也可以和泡菜一起吃，或者配点萝卜，小肉片和面条。

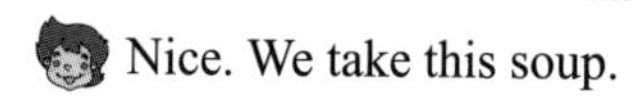

Nice. We take this soup.

不错，我们就要这个汤吧。

开胃词组

go with 伴随，与……相配

stick to 紧跟，跟随

warm up 使暖和

as well 也，还有

鲜美单词

interesting adj. 有趣的，令人感兴趣的

prepare v. 准备（食物），做（菜）

habit n. 习惯，习性

ox n. 牛，公牛

milky adj. 乳状的，乳白色的

aroma n. 芳香，香味

meaty adj. 多肉的

broth *n.* 肉汤

roof *n.* 口腔顶部，上腭

brisket *n.* 胸肉，牛胸肉

Conversation 2

Could you tell me more about Korean cuisine?	你能跟我讲讲更多韩国菜方面的知识吗?
OK. Today I will introduce fantastic Korean soup to you.	好的，今天我给你介绍超级棒的韩国汤。
Soup? I like to drink soup, go ahead.	汤？我喜欢喝汤！继续讲吧。
Well, soup is an integral part of Korean cuisine. Soup is usually served as part of the main meal along with rice and other banchan which comprise of side dishes.	嗯，汤是韩国菜不可缺少的一部分。汤作为主菜的一部分，跟米饭和小菜一起吃。
Ah…it sounds a bit different from the western soup. In western countries, it is served as an appetizer.	噢，听起来跟西餐不一样啊。在西方，汤是一种开胃食品。
Oh, yeah. Most Korean soup can be classified into two primary groups, guk and tang. The former is watery and served at homes while the tangs are available at the commercial eateries. Meat, shellfish and vegetables are the usual ingredients of the Korean soups.	哦，是的。韩式汤大多分为两种：大酱汤和果酱汤。前者大酱汤可以在家自制，果酱汤在餐馆可以买到。肉类，贝类和蔬菜是韩式汤的主要材料。
You must know a lot of Korean soup recipes.	你肯定了解许多韩式汤的菜谱。
I know a few, such as gomguk, naengguk, malgeunguk, tojangguk…	我知道几个菜谱，牛杂汤、凉汤、清酱汤、大酱汤……

Wow, you are great.

哇，你真棒！

Thank you.

谢谢！

开胃词组

along with 和……一起，随着

comprise of 组成，构成

different from 不同……，异于……

be classified into 分为……

鲜美单词

cuisine n. 烹饪；菜肴

introduce v. 介绍，引进

fantastic adj. 极好的，极秒的

integral adj. 完整的，积分的

comprise v. 由……组成，由……构成

western adj. 西方的，欧美的

appetizer n. 开胃品，开胃菜

classify v. 分类，归类

primary adj. 首要的，主要的

former pron. 前者，前一个

watery adj. 含水的，水分过多的

commercial adj. 商业的，商业化的

Korean seaweed soup is a definitely healthy and delicious way to incorporate seaweed into your diet. Traditionally, Koreans would serve this on birthdays and after giving birth. They say that the seaweed helps the new mother to heal.

韩式海带汤是一道健康的美食，让海带成为你的盘中餐。按照韩国的传统，这道菜是庆生和孕妇分娩之后的特色菜。他们说海带有助于新妈妈们恢复身体。

It is a long tradition in Korea that new mothers eat seaweed soup for certain period of time after childbirth since this soup would bring them strength. Seaweed soup has about 40 kinds of minerals, DHA, vitamins, iodine and other nutrients. It has 200 times more calcium than rice, 25 times more than spinach and 13 times more than milk. Seaweed soup is known to be special food for new mothers who are recovering from the loss of blood during childbirth by providing healthy doses of all the nutrients.

妇女生完孩子，需要吃一段时间海带汤，来恢复她们的体力，这已经是韩国的一个老传统了。海带汤内含40多种矿物质、DHA、维生素、碘及其他营养物质。它比米饭的含钙量高200倍，比菠菜高25倍，比牛奶高13倍。海带汤是产妇们的特色食物，帮助她们恢复分娩过程中因失血而丢失的所有营养物质。

It’s also a traditional birthday dish originating from the Goryeo Dynasty. This custom implies that you should not forget how grateful you should be to your mother for her giving birth to you by eating seaweed soup on the morning of your birthday.

自高丽王朝起，海带汤已经是韩国人生日庆祝的传统菜。生日那天的早晨喝一碗海带汤的传统，能够让人们在庆祝生日的同时，感恩赐予生命的母亲们。

Korean Noodles
韩国面

舌尖美食词汇

bibimguksu 拌面	Korean stir fried noodles 韩式炒面
guksu jangguk 面条酱汤	janchi guksu 喜面
japchae 什锦炒菜	makguksu 爽口鸡汤荞麦面
vegetarian ramyeon 蔬菜拉面	jjolmyeon 筋面
kalguksu 刀切面	pumpkin noodles 南瓜面

舌尖美食句

1. Do you like bibimguksu? 你喜欢吃拌面吗？
2. Korean stir fried noodles are a must for my dinner. 韩式炒面是我晚餐必吃的菜。
3. In Korea, a traditional noodle dish made with a clear broth is called Guksu Jangguk. 韩国一道用清汤做传统面食叫面条酱汤。
4. Janchi guksu is usually served with fried egg or zucchini on top. 喜面通常在上面放着炒蛋或西葫芦。
5. Japchae is a kind of mixed vegetable with noodles. 什锦杂菜是各种蔬菜和面条的杂烩。

6. I love makguksu, especially the tasty clear chichen soup.	我喜欢吃爽口鸡汤荞麦面，尤其里面的鸡肉清汤。
7. Vegetarian ramyeon is flavorful.	蔬菜拉面很好吃。
8. Jjolmyeon is similar to bibim naengmyeon but the noodles are more chewy.	筋面与拌面相似，但面条更耐嚼。
9. Kalguksu are handmade, knife-cut noodles which are served in large bowls.	刀切面是手工制作，用刀切好的面条，放在大碗里食用。
10. Pumpkin noodles are my favorite staple food.	南瓜面是我喜欢吃的主食。

舌尖聊美食

Conversation 1

How can I help you, sir?	需要帮忙吗，先生？
What kind of staple food do you serve here?	你们这里有什么主食吗？
We have rice and noodles. The rice is served in large bowls, which could be suitable for five or seven people. And we have different types of noodles in small bowls. Which one would you like?	我们有米饭和面条。米饭是大碗的，够5到7个人吃。我们还有很多种面条，小碗装的。您想要哪个？
Well, what kind of noodles do you provide?	嗯，你们有哪种面条？
These noodles are all authentic Korean noodles. These tasty noodles are all handmade right after you order. We also have Korean stir fried noodles, naengmyeon and kalguksu.	这些面条都正宗的韩国面条。这些美味面条都是现时手工制作的。我们还有韩国炒面、冷面和刀切面。

So, a small bowl of stir fried noodles and a bowl of kalguksu, please.

那么，我们来一小碗炒面和一碗刀切面吧。

OK. They will be ready to serve right away.

好的。马上就给您准备好。

开胃词组

staple food 主食

right after 刚好在……之后

be suitable for 适合，适宜

right away 马上，立刻

鲜美单词

staple adj. （食物、产品等）主要的，主打的

noodle n. 面条

bowl n. 碗

suitable adj. 合适的，适宜的

provide v. 提供，供给

authentic adj. 真的，正宗的

tasty adj. 美味的，可口的

handmade adj. 手工做的

order n. 点菜，订单

serve v. 提供，端上

Conversation 2

Eve, is the naengmyeon hot?

伊芙，冷面辣吗？

I think so. It's usually served with red pepper paste and garlic. If you're not into the spicies, we have other selections. How about japchae? It has a mild flavor.

我认为是的。一般冷面里面都放红辣椒和大蒜的。如果你不喜欢吃辣的，我们还可以选其他的。什锦炒菜，怎么样？这个味道还是很清淡的。

Oh, that's great. Could we choose the vegetables we prefer?

哦，那好。我们可以选择放自己喜欢的蔬菜吗？

Yeah. You can choose Korean cabbage and parsley in this season, as those have the best flavor now.

是的，你可以选韩式大白菜，还有香芹，这个季节吃，味道最好。

Good. What about you? You love the spicies, huh?

好，你呢？你喜欢吃辣的，是吧？

Definitely! Spicy buckwheat noodles are my favorite.

当然啦！辣味荞麦面是我的最爱。

OK, let's make the order now.

好的，我们开始下单吧。

开胃词组

red pepper paste 红辣椒酱

be into 喜欢，迷上

what about 怎么样

make the order 下菜单，点餐

鲜美单词

pepper n. 辣椒

paste n. 面团；糨糊

garlic n. 大蒜

selection n. 选择

mild adj. 轻微的，不浓的

cabbage n. 圆白菜，大白菜

parsley *n.* 西芹
season *n.* 季节
buckwheat *n.* 荞麦

舌尖美食文化

South Korea may produce probably the best instant noodles, but for me it's the home made dishes which are the best. Noodles are a staple in South Korea similar to rice. When you don't fancy rice, have a dish with noodles instead!

韩国也许是生产方便面最好的国家，但对于我来说，我觉得韩国的自制面食才是最好的。面条和米饭都是韩国的主食。如果你不喜欢吃米饭，吃面条也不错。

Japchae noodles are unique and famous. The noodles are made from sweet potato starch and called glass noodles, these have a very different texture and taste to ordinary noodles.

什锦炒菜因独特而闻名四方。这种面条是土豆淀粉做的，也叫玻璃面条，与普通面条不同，它有独特的质地和味道。

Sundae bokkeum is a noodle dish which uses the Korean Sundae as its main ingredient with added vegetables. It is hot and spicy and the sausage makes it a very filling meal.

炒面肠是用韩国血肠做的，主料是蔬菜。这道菜不辣，但酱汁让这道菜很美味。

Jeyuk bokkeum is also a famous Korean noodles dish. It's a simple and quick spicy dish. The pork gives a great texture and makes sure this dish is very filling.

辣炒猪肉也是一道著名的韩式面食。这道菜简单而辛辣，里面的猪肉味道鲜美，让整道菜吃起来意犹未尽。

读书笔记

Part 9

Start Eating
开始用餐

UNIT 01 Western Table Manners 西餐礼仪

舌尖美食词汇

table manners 餐桌礼仪	**hold the fork** 手拿餐叉
excuse me 失陪一下	**pay tips** 付小费
pick up 拾起	**keep the table clean** 保持桌面干净
chew your food 嚼食物	**eat noisily** 大声地吃
fork on the left 餐叉放在左手边	**pass along** 传递

舌尖美食句

1. Table manners are fundamental to every public dining situation.	公共用餐场合中，餐桌礼仪是非常重要的。
2. Which hand should I use to hold the fork?	应该用哪只手来拿叉子啊？
3. Don't slurp your soup.	喝汤不要发出声音。
4. Please keep the table clean.	请保持桌面干净。
5. Chew with your mouth closed.	闭着嘴巴嚼食物。

6. Excuse me for minutes.	我失陪一会儿。
7. Don't put your elbows on the table.	不要把手肘支在餐桌上。
8. Excuse me. I'll just get the phone.	对不起，我去接个电话。
9. Remember to pay tips after eating.	记得饭后给小费。
10. It is impolite to reach over someone to pick up food or other items.	跨过别人而去取自己想要的食物是非常不礼貌的。
11. The napkin should be left on the seat when leaving temporarily.	临时离开餐桌的时候，应该把餐巾放在椅子上。

舌尖聊美食

Conversation 1

How to handle a napkin at western dinner?

西式晚宴该怎么摆放餐巾？

Generally, when you are seated, put your napkin on your lap. But at a formal dinner, you should wait until the hostess picks up her napkin, then pick yours up and lay it on your lap. Sometimes a roll of bread is wrapped in it; if so, take it out and put it on your side plate.

一般而言，当你入座后，应把餐巾放到腿上。但是在正式宴会上需要等女主人首先拿起餐巾搁在她自己腿上后，你才拿起自己的餐巾放到腿上。有时候，餐巾上面已经放了面包,应该把面包放到配菜盘子里。

How do I know when to start eating as a guest at dinner?

作为客人，我怎么知道什么时候开始吃呢?

Dinner usually begins with soup. The host or hostess will ask guests to begin with the soup after they have been served.

晚宴一般都从喝汤开始。汤分到每个客人手里，主人或女主人会叫大家一起喝汤，晚宴正式开始了。

Part 9 开始用餐

 What about the table setting?

桌子餐具的摆放呢?

 Well, a plate will be put in the middle of your place. Forks are on your left side and nives are on your right side. Dessert cutlery is at the twelve o'clock of your plate, and bread cutlery at eleven o'clock.

嗯，你的桌子位置中间会有一个餐盘，餐叉放在左手边，餐刀放在右手边。点心餐具放在餐盘的12点钟的位置，面包餐具放在11点钟的位置。

 How about the glasses?

杯子怎么摆放呢?

 Water glass, red wine glass and white wine glass are palced at one clock.

水杯、红葡萄酒杯、白葡萄酒杯分别放在1点钟的位置。

 OK. I got it. Thank you.

好的，我明白了。谢谢您!

 You're welcome.

不客气。

开胃词组

pick up 捡起，拾起

take out 取出，拿出

begin with 以……开始

white wine 白葡萄酒

鲜美单词

napkin *n.* 餐巾，餐巾纸

western *adj.* 西方的，欧美的

generally *adv.* 一般地，通常地

seat *v.* 就座，坐下

formal *adj.* 正式的，庄重的

hostess *n.* 女主人

lay *v.* 放置，安放

wrap *v.* 包，缠绕
guest *n.* 客人，宾客
fork *n.* 餐叉

Conversation 2

I'm going to have dinner with Jason at a western restaurant this evening, and I'm worried.	今天晚上要去和詹森吃西餐。我有些担心。
Why?	为什么？
I'm not familiar with the table manners.	我对餐桌礼仪不太了解。
Well, just follow your hostess and remember not to make any noise, especially with soup, coffee, water and other drinks. They usually use the knife and spoon in their right hand, and the fork in their left hand.	嗯，你只要跟随主人并注意吃饭时不要出声，尤其是喝汤、咖啡、水和其他饮料的时候。他们一般用右手使用餐刀和勺子，左手使用餐叉。
Of course, I know that. But what should I do if I drop a spoon or fork on the floor?	当然，我知道那些。但万一汤匙或叉子掉地上了，我该怎么办？
Don't worry about that. If ever that happens, just pick it up.	不用担心。如果真的掉了东西，捡起来就行了。
Is there anything else I should take care with?	还有没有其他我该注意的事？
I think so. When you sit down straight, put the napkin on your lap. No food in your mouth when you're talking.	是的，当你坐下来之后，应该把餐巾纸放腿上，说话的时候，嘴里不要含着食物。
I see. Thank you.	我知道了。谢谢您！
You are welcome.	不用客气。

开胃词组

be familiar with 对……熟悉

make noise 弄出噪声，出声

worry about 担心

if ever 如果真的……

take care 小心，当心

鲜美单词

familiar adj. 熟悉的，通晓的

manners n. 礼仪，礼节

noise n. 喧闹声，嘈杂声

drink n. 饮料，酒精饮料

spoon n. 匙，小勺子

happen v. 发生

straight adv. 笔直地，径直地

lap n. 大腿

mouth n. 口，嘴

舌尖美食文化

If you have English and American friends, you will notice a few differences in their customs of eating. For the main or meat course, the English keep the fork in the left hand, point curved downward, and bring the food to the mouth either by sticking the points onto it or in the case of soft vegetables, by placing it firmly on the fork in this position with the knife. Americans carve the meat in the same position, then lay down the knife and taking the fork in the right hand with the point turned up, push it under a small piece of food without the help of the knife and bring it to the mouth right-side-up.

如果你有英国和美国朋友，你会发现他们的饮食习惯会有些不同。上主菜时，英国人习惯把餐叉放在左手边，叉子朝下，把食物送到嘴里。当吃松软的蔬菜时，餐叉叉口朝上用餐刀把食物放到餐叉上。美国人习惯左手拿餐叉，叉口朝下，用餐刀切好肉，把餐刀放下，换过来用右手拿餐叉，开口朝上，把食物送到嘴里吃。

On formal dining occasions, it is acceptable to take some butter from the butter dish with your bread knife and put it on your side plate. Then butter pieces of the roll using this butter. This prevents the butter in the dish getting full of bread crumbs as it is passed around. Knives should be used to butter bread rolls but not to cut them — tear off a mouthful at a time with your hands.

在正式宴会场合，吃面包时，拿你的面包餐刀，去取一点黄油，然后把黄油放在自己的小盘子边上，这样做是可以的。然后把黄油涂在你的小面包片上。在传递的时候，黄油上很容易沾满面包屑，要多加注意。餐刀是用来取黄油的，不是用来切面包的，应该是用手每次从面包上撕下一小块来吃。

读书笔记

UNIT 02 Drinking Wine 美酒交杯

舌尖美食词汇

cardinal cocktail 卡蒂娜鸡尾酒	**Belgain orange blossom** 比利时橙花酒
cider 苹果酒	**martini** 马提尼
Gin and Tonic 金汤力鸡尾酒	**rum** 朗姆酒
champagne 香槟酒	**Tequila Sunrise** 龙舌兰日出
margarita 玛格丽特酒	**whiskey** 威士忌

舌尖美食句

1. It's really simple to make cardinal cocktail. 制作卡蒂娜鸡尾酒真的很简单。
2. Would you like to try Belgain orange blossom? 你想试一试比利时橙花酒吗？
3. This stout makes me tipsy. 这个烈酒令我喝醉了。
4. What kind of spirits do you like? 你喜欢喝哪种烈酒？

5. What's the difference between cider and martini?	苹果酒和马提尼有什么区别？
6. Help yourself to gin and tonic.	来点金汤力鸡尾酒吧。
7. You're staggering. You should stop drinking more rum.	你喝得摇摇晃晃的。不要再喝朗姆酒了。
8. This champagne tastes pretty mellow.	这杯香槟酒喝起来很香醇。
9. I like to drink Tequila Sunrise, a kind of cocktail which is made with tequila and orange juice.	我喜欢喝龙舌兰日出，一种龙舌兰和橙汁调制的鸡尾酒。
10. Why don't you have a glass of margarita? It's sweet and tangy.	要不来点玛格丽特酒？这种酒甜甜的，味道浓厚。
11. Can I have shot of whiskey on the rocks?	可以给我来一杯加冰块的威士忌吗？

舌尖聊美食

Conversation 1

Would you like a drink?	你想点什么喝的吗？
Why not? I would like a glass of Gin and Tonic.	好的，为什么不呢？我想来一杯金汤力。
I like Gin and Tonic, too. But tonight I think I'll have the special one.	我也喜欢喝金汤力。但是今晚我想喝那种特色酒。
Excuse me?	你说什么？

Oh, it's just a type of shooter. It's got Amaretto, Kahlua and Irish cream in it.	噢，那是一种鸡尾酒，里面有杏仁酒、香甜咖啡酒和爱尔兰奶酒。
What does it taste like?	喝起来怎么样？
It's quite sweet. Do you want to try one?	相当甜。你要不要尝一尝？
No, thanks. I'm not big on liquors.	不了，谢谢！我不太喜欢烈酒。
OK. Our wines are prepared. Here is your Gin and Tonic.	好吧。我们的酒准备好了，这是你的金汤力。
Cheers.	干杯！
Cheers.	干杯！

开胃词组

why not 为什么不呢

a glass of 一杯

be (not) big on sth. 很喜欢，喜爱

鲜美单词

drink *n.* 酒，饮料

glass *n.* 玻璃杯

gin *n.* 杜松子酒

tonic *n.* 奎宁水

shooter *n.* 射手，枪手

amaretto *n.* 意大利苦杏酒

liquor *n.* 酒，烈性酒

Conversation 2

Hi, could you bring us your drink list?	嗨！您能给我们拿酒水单来吗？
Sure. Here you are.	可以。给您！
Wow, there're so many drinks here. Could you recommend today's special?	哇！有很多酒可以点。今天您有什么特别推荐的吗？
Sidecar and Belgian orange blossom are very popular. Both of them are cocktails.	边车白兰地和比利时橙花酒都很受欢迎。这两种酒都是鸡尾酒。
Linda, what would you like to drink?	琳达，你想喝点什么？
I prefer drinking champagne.	我更喜欢香槟酒。
Oh, yeah, we have champagne cocktail. It's made with Angostura bitters, champagne, brandy and a maraschino cherry as a garnish. Add dash of Angostura bitter onto sugar cube and drop it into champagne flute. Add cognac followed by gently pouring chilled champagne. Garnish with orange slice and maraschino cherry.	噢，对了，我们有香槟鸡尾酒。用安格斯特拉苦酒、香槟酒、白兰地调制的，并用马拉斯奇诺樱桃装饰。先在少量的安格斯特拉苦酒里放点糖块，再倒进长形香槟酒杯，往里加点干邑白兰地，然后慢慢倒进香槟。最后用橙子片和马拉斯奇诺樱桃装饰。
I have never heard of it. I think I would like to have a try.	我从来没听说过这种酒。我想我可以试一试。
OK, we take this order.	好的，我们就点这个。

开胃词组

be popular 流行，受欢迎

both of 两者都

hear of 听说，听过

have a try 尝试

鲜美单词

bring v. 带来，引来

recommend v. 推荐，劝告

blossom n. 花，花簇

cocktail n. 鸡尾酒

champagne n. 香槟酒

Angostura n. 安格斯特拉苦酒

bitter adj. 苦的，苦味酒

brandy n. 白兰地

garnish n. 装饰

dash n. 少量

cognac n. 干邑白兰地

舌尖美食文化

Cocktail is an alcoholic drink which was originally a mixture of spirits, sugar, water and bitters. Now, it is simply a mixture of drinks which contains three or more ingredients with at least one that must be a spirit.

鸡尾酒是一种由烈酒、糖、水、苦味酒混合调制而成的酒精饮料。现在鸡尾酒是一种包含三种或多种食材，至少一种烈酒的混合饮料。

There are as many stories behind the origin of the name cocktail as there are behind the creation of the first Margarita or the Martini. A popular story behind the cocktail name refers to a rooster's tail (orcock tail) being used as a Colonial drink garnish. There are no formal references in written recipes to such agarnish.

关于鸡尾酒起源的故事，与玛格丽特或者马提尼背后的故事一样

多。普遍认为鸡尾酒在美国殖民属地时期因以鸡尾毛做装饰而得名。书上没有记载如何做这种装饰。

Although there are hundreds and hundreds of different cocktails, many of them may be grouped together in “families”, based on a common recipe. For example, although traditional Manhattansare based on whiskey, other members of the Manhattan family may contain southern comfort, brandy or rum in its place.

尽管有数百上千种的鸡尾酒，但是它们因相似的配制方法而属一大类，例如，尽管传统的曼哈顿鸡尾酒以威士忌为基酒，而其他的曼哈顿鸡尾酒有的以金馥力娇酒、白兰地酒、朗姆酒作为基酒。

读书笔记

UNIT 03

Family Dinner
家庭用餐

舌尖美食词汇

grilled lamb chop 烤羊肉	**bread rolls** 面包卷
shepherd's pie 农家土豆肉馅饼	**trifle** 水果松糕
chicken cacciatore 意式鸡肉蔬菜煲	**chutney** 酸辣酱
seafood soup 海鲜汤	**barley tea** 大麦茶
kipper 腌鱼	**salami** 意式香肠
bangers and mash 肉泥土豆泥	**tamales** 墨西哥玉米粉蒸肉

舌尖美食句

1. Would you like to try this grilled lamb chop?	你要尝尝这个烤羊肉吗？
2. Could you pass me the bread rolls?	你能给我递一下面包卷吗？
3. You have eaten so much shepherd's pie.	你吃了太多的农家土豆肉馅饼。
4. This trifle just fits your appetite.	这个水果松糕正合你的胃口。

5. Chicken cacciatore is a tasty dish that could feed the whole family.

意式鸡肉蔬菜煲是一道美味菜肴，可供全家人食用。

6. Would you like some more chutney?

你想再来一些酸辣酱吗？

7. We'd better start with this seafood soup.

我们最好先喝这个海鲜汤。

8. Barley tea is typically served after the meal.

大麦茶通常在饭后饮用。

9. We often have kipper instead of salami.

我们常吃腌鱼，而不常吃意式香肠。

10. What's wrong with the bangers and mash? It has no taste at all.

这个肉泥土豆泥怎么回事？一点也没有味道。

11. Tamales are fabulous for family gathering.

墨西哥玉米粉蒸肉是家庭聚餐的美味佳肴。

舌尖聊美食

Conversation 1

Mom, I'm very curious about the term "Sunday Roast". Do British people only eat roast meat on Sunday? What is the origin of this tradition?

妈妈，我一直很好奇星期日烤肉的由来，英国人是在星期天吃烤肉吗？这个传统的起源是什么？

Well, honey, listen. It's called "Sunday Roast" because people go to church on Sundays, and the meat can be put in the oven to cook before the family goes to church and be ready to eat when they return at lunchtime.

嗯，宝贝，听着，星期日烤肉这个名称的由来是因为人们星期天要去教堂做礼拜，去教堂之前，他们把肉放进炉子里烤，等他们回来时，就有午餐吃了。

I see. Mmm… the roast pork is yummy. I think it's also good for the dinner. Can we cook it for our dinner today?	我明白了。嗯……烤猪肉很好吃，我觉得晚饭吃这个也很好啊。我们可以做烤肉作为我们今天的晚餐吗？
Sure. And we can have it as a surprise for your father.	当然。我们可以给你爸爸来一个惊喜。
Wonderful. What's the brown, crispy thing in the picture of this recipe?	好极了！菜谱图片里这个棕色的，看起来脆脆的是什么东西？
That's crackling, the skin of the pork. Traditionally, roast pork is served with crackling, onion stuffing, apple sauce, and English mustard.	那是脆烤猪皮，也就是猪皮。传统上，烤猪肉要加上脆皮、洋葱填料、苹果酱和英式芥末的。
Any other meats for Sunday Roast?	星期天烤肉还吃什么肉呢？
Beef, lamb, and sometime a whole chicken.	还有牛肉、羊肉，但有时是一整只鸡。
Wow, I hope we could eat all these tasty food tonight.	哇！我希望我们今晚能够吃遍这些美食。
Oh, in that case, I would need your help.	噢，如果那样的话，我得需要你的帮助才行。
Sure. I love to.	当然，我愿意做帮手！

开胃词组

be curious about 对…… 好奇

the origin of ……的起源

at lunchtime 午餐时间

as a surprise 作为一个惊喜

in that case 如果那样的话

鲜美单词

curious *adj.* 好奇的，稀奇的

origin *n.* 起源，根源

honey *n.* 宝贝，可爱的人

church *n.* 教堂

return *v.* 回来，返回

roast *v.* 烘，烤

surprise *n.* 惊喜，惊奇

crispy *adj.* 脆的，酥脆的

crackling *n.* （烤猪肉的）脆皮

stuffing *n.* 填充物，填料

mustard *n.* 芥末

Conversation 2

Dinner time!

吃晚饭啦！

Wow, mom, you prepared such a great dinner.

哇，妈妈。您准备了这么丰盛的晚餐啊。

Thank you. These are just some of my specialties.

谢谢！这只是一些我的拿手菜。

Look, daddy, this is your favorite fried chicken wings. And that's my lovely Steak Diane. Oh, chocolate mousse, souffle… all the best French cuisine. I love you, mom!

你看，爸爸，这是您最喜欢的炸鸡翅，还有我最喜欢的戴安娜牛排。哦，还有巧克力慕斯，蛋奶酥……全是最好的法式菜肴。我爱你，妈妈！

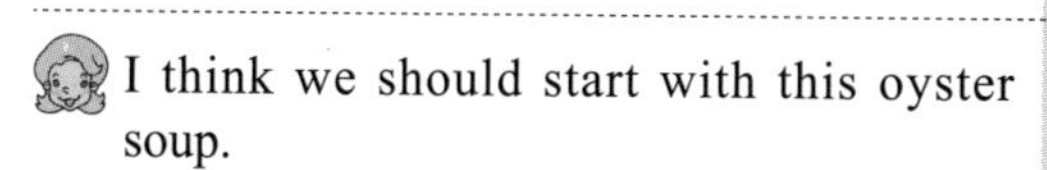

I think we should start with this oyster soup.

我们应该先喝点牡蛎汤。

Yes, that's a must of your delicacy. Could you pass me the bowl and I serve this beautiful soup for you?	是的，这是您必做的一道菜啦。您能递给我汤碗吗？我给您盛汤。
Oh, very nice, sweetheart. Thank you.	哦，好的，宝贝。谢谢！
Well, I just drink a little bit of soup. I need more room for the main dish and desserts.	嗯，我只喝一点汤就可以啦。我还要留着肚子吃主菜和点心呢。
Do you like another piece of steak, Tina?	你再来点牛排吗，蒂娜？
No. Thanks, dad. I'm going to eat the tasty chocolate mousse.	不了，谢谢您，爸爸。我打算吃那美味的巧克力慕斯了。
Haha, how about drinking some champagne for this wonderful dinner?	哈哈，我们来喝点香槟庆祝这美味的晚餐，怎么样？
Great. Cheers.	好啊，干杯！
Cheers.	干杯！

开胃词组

start with 以……开始

main dish 主菜

how about 怎么样，如何

鲜美单词

dinner *n.* 正餐，晚餐

prepare *v.* 准备，预备

specialty *n.* 特色菜，招牌菜

steak *n.* 牛排

souffle *n.* 蛋奶酥

oyster *n.* 牡蛎

delicacy *n.* 美味，佳肴

champagne *n.* 香槟酒

You might be wondering what's for dinner. I have some quick dinner ideas:

你也许在想晚餐吃什么呢。我这里有一些方便快捷的晚餐推荐：

Pasta — Pasta is known for its simplicity and can be prepared in minutes! All you need to do for a simple classic pasta is to boil the pasta, add some tomato sauce, dried herbs (optional), some cheese and your filling meal is ready! To bring up the nutritional quotient, you can add in some stir-fried vegetables, Italian sausage, shredded chicken, etc. into the pasta. If you are bored of having the usual spaghetti, try preparing a frittata, which is also easy and a different way of eating spaghetti.

意大利面——意大利面制作简单，数分钟就可以搞定！要做一份简单经典的意大利面，你只需把意大利面煮熟，然后往里加一些番茄酱、香草（可选）和一些奶酪，这样，你的晚餐就准备好了！如果要让意大利面更有营养，只需加一点爆炒蔬菜、意大利酱、鸡丁等等。如果不喜欢普通的意大利细面条，可以做个意大利煎蛋饼和意大利面一起吃，同样很简单，与众不同的吃意大利面的方式。

Stir fry — Stir fry is so simple to prepare, yet form a delicious meal. All you need is some soy sauce, cornstarch, seasoning and meat of your choice. Stir them up in a wok and have your lovely stir fry for dinner over a bed of rice or noodles. These are great ways of feeding vegetables to kids. If the kids dislike a particular vegetable, grate them or slice them really small, so they can't identify them in the stir fry. For a delicious stir fry, all you need is some creativity!

干煸小肉丁——干煸小肉丁简单而美味可口。只需准备一些酱油、淀粉、佐料和肉类。把这些食材放锅里爆炒，然后与米饭或者面条食用。这是一种让孩子吃蔬菜的好方法。如果孩子们不吃某种蔬菜，可以把这种蔬菜切小一点，好让他们辨认不出来。味美可口的干煸小肉丁需要你的创意！

Risotto — If you're looking for a quick, but comforting dinner idea, then risotto is the perfect dish! Always keep arborio, carnaroli or any other Italian risotto in your kitchen cupboard. To prepare a simple risotto, you need risotto rice, chicken broth (any broth), dry white wine, cheese and seasonings. Since the broth has to be added a ladle at a time, it will take 25-30 minutes to prepare the risotto. For a more appealing and filling risotto, add mushrooms, chicken, sausage, shrimp, peas, pumpkin, etc.

意大利肉汁烩饭——如果你在寻找方便快捷、美味的晚餐，意大利调味饭可能就是你的完美之选！平时厨房橱柜里准备一些意大利圆粒米、卡纳罗利米或其他调味饭用米，就会很方便。做一个简单的意大利肉汁烩饭，你只需准备烩饭用米、鸡汤（或任何肉汤）、干白葡萄酒、奶酪和佐料。因为每次要添加一勺肉汤，做一个意大利肉汁烩饭得花25至30分钟。要做更丰盛诱人的意大利肉汁烩饭，可以再加点蘑菇、鸡肉、酱油、虾仁、豌豆或南瓜等。

Holiday Party 节日派对

舌尖美食词汇

New Year Eve 除夕夜	**Christmas party** 圣诞节派对
Happy New Year! 新年快乐！	**Thanksgiving** 感恩节
roast turkey 烤火鸡	**pudding** 布丁
panettone 潘娜托尼蛋糕	**candy** 糖果
pumpkin pie 南瓜派	**apple tart** 苹果塔

舌尖美食句

1. Roast turkey, smashed potatoes and a pudding are cooked for the Thanksgiving supper.	烤火鸡、土豆泥和布丁是感恩节的食物。
2. I'm very excited to eat panettone on the New Year Eve.	我真的很兴奋在除夕夜吃潘娜托尼蛋糕。
3. Thank you very much for the wonderful gift.	谢谢你给我这么好的礼物。
4. It's quite easy to make melegatti.	莫丽塔蛋糕的制作相当简单。

5.	I am going to make a pumpkin pie.	我要做南瓜派。
6.	I love to eat chocolate and candy at Easter time.	复活节我喜欢吃巧克力和糖果。
7.	This is the best apple tart I've ever had.	这是我吃过的最好吃的苹果塔。
8.	Thank you for your coming to our Christmas party.	谢谢你能来参加我们的圣诞节派对！
9.	Happy New Year!	新年快乐！
10.	These mince pies are tasty.	这些百果馅饼真好吃！

舌尖聊美食

Conversation 1

Mrs. White, thank you for inviting me to this party. I was always looking forward to seeing what traditional Christmas dinner is like.	怀特女士，谢谢您邀请我来参加派对！我一直想知道传统的圣诞节晚餐都吃什么。
You're welcome, Lily. You will see it. Well, dinner is ready. Let's go to the dinning room, please.	别客气，莉莉！你会看到的。哦，晚餐准备好了。我们去餐厅吧。
OK. Wow! What a feast! That's a huge chicken.	好的。哇！好丰盛的晚餐啊！这只鸡好大呀。
That's turkey, Lily. Roast turkey is what we usually have for the Christmas dinner. My husband will cut the turkey for us, and it's usually served with some gravy on top.	那是火鸡，莉莉。我们圣诞节晚餐通常吃烤火鸡。我的丈夫会把它切好的，还有火鸡肉通常淋着肉汁一起吃。

Mmm… the turkey is very tasty. I think I would like some more.

嗯，火鸡肉真美味！我想再吃一点。

Haha… please save some room for the Christmas pudding, my husband's specialty.

哈哈，注意留一点肚子吃圣诞节布丁啊，这可是我丈夫的招牌菜。

I love pudding. I won't miss it.

我喜欢布丁，我不会错过的。

开胃词组

thank you for doing sth. 谢谢你做某事

look forward to 期待

dinning room 餐厅，吃饭的地方

on top 在上面，放在上面

鲜美单词

invite v. 邀请，请求

traditional adj. 传统的，习俗的

Christmas n. 圣诞节

feast n. 盛会，盛宴

huge adj. 巨大的，庞大的

roast v. 烘，烤，焙

turkey n. 火鸡

gravy n. 肉汁，肉卤

pudding n. 布丁

room n. 空间，余地

miss v. 漏掉，错过

Conversation 2

Happy Thanksgiving Day, Ben!	感恩节快乐，本!
Happy Thanksgiving Day! Have you got any plan?	感恩节快乐！你有什么计划吗?
Not really, all my friends went home. I will be alone here.	没有，我的朋友都回家了。我要一人在这里度过了。
Why don't you come to my family's dinner party? My parents will very glad to meet you.	你为什么不来我家参加晚宴派对呢？我父母会很高兴认识你的。
That's great. I'd love to.Thank you.	那太好了。我很乐意去。谢谢你。
Wonderful. I'm excited we can enjoy this great Thanksgiving Day together.	那太棒了，我们一起度过美好的感恩节，我真的很兴奋。
Could you tell me something about it?	你能给我讲讲感恩节吗?
It's usually a big family reunion.We always have a lot of delicacy to celebrate Thanksgiving, like turkey and stuffing, but sometimes we have ham instead.We also have yams, corn, carrots and fruit pies or pumpkin pies.	感恩节通常是家庭大聚会。我们会做很多美味佳肴，比如说火鸡，火鸡馅，不过我们有时也吃火腿来代替火鸡。我们还吃洋芋、玉米、胡萝卜、水果或南瓜派。
That sounds like a big feast, just like our reunion dinner on Spring Festival. I am looking forward to the big thanksgiving dinner now!	听起来真像个盛大的宴会，就像我们的春节团圆饭。我很期待这盛大感恩节晚餐!
Me too. I can't wait to see my family.	我也是。我都等不及想见家人了。

开胃词组

be glad to meet you 很高兴认识你

family reunion 家庭团聚

wait to do sth. 等着做某事

鲜美单词

Thanksgiving n. 感恩节

plan n. 计划，打算

alone adj. 单独的，孤独的

reunion n. 团聚，团圆

excited adj. 兴奋的，激动的

turkey n. 火鸡

yam n. 山药，洋芋

instead adv. 作为代替，反而

pumpkin n. 南瓜

stuffing n. 填充物，馅料

festival n. 节日

舌尖美食文化

Thanksgiving dinner is one of the most important meals of the year. It is a chance to spend some quality time with your family members and friends, plus for cooks. It is the time to prepare traditional comfort food.

感恩节晚宴作为一年当中最为重要的晚宴之一。这一天，准备丰盛的传统食物，与家人或朋友一起庆祝节日，是一次令人难忘的美好时光。

Thanksgiving is not Thanksgiving without the turkey. A relatively new technique in cooking turkey is deep-frying the bird before roasting it, giving it crisp, golden brown skin. Keep the turkey moist by regularly basting it with its own drippings, plus don't forget the stuffing for additional flavor and moisture inside the bird's cavity. A secondary main dish during Thanksgiving is a leg of ham. For those who are not fond of turkey, a leg of oven-roasted ham with honey Dijon sauce is an excellent alternative.

感恩节不能没有火鸡。一种最新的烤火鸡的方法就是在烤火鸡之前，先把火鸡炸一遍，让它的外皮金黄而又酥脆。为保持火鸡湿润，需在鸡肉上涂抹其本身的油脂，别忘了填料作为调味品，以保持火鸡内部湿润。感恩节的第二道主菜当属火腿。那些不喜欢吃火鸡的人，可以享用带着蜂蜜酱的火腿。

Pumpkin pies, whether store-bought or homemade, cap most Thanksgiving meals. Add some pecans to the pumpkin pie's crust for a crunchier take on the classic. You can also try making a pumpkin cheesecake, instead of the more traditional pumpkin pie. Another classic American dish is the apple pie, which can be enhanced by pouring some caramel sauce directly on top of the pie just before serving it. Add some crunch to the pie by toasting some walnuts or almonds and sprinkling them on top of the creation.

南瓜派，无论是购买的还是家庭自制的，都是感恩节的点睛之笔。用山核桃放在南瓜派外层皮上，吃起来脆脆的，很经典。你可以试试做南瓜奶酪饼，作为传统南瓜派的替代品。另一道经典美式甜品就是苹果派，吃的时候，淋上一点焦糖更美味。制作的时候，在苹果派上面放一点胡桃仁或杏仁会更脆。